R

D1179364

WHISPERS OF THE DEAD

www.**rbooks**.co.uk

Also by Simon Beckett
featuring David Hunter

THE CHEMISTRY OF DEATH
WRITTEN IN BONE

For more information on Simon Beckett and his books, see his
website at www.simonbeckett.com

WHISPERS OF THE DEAD

SIMON BECKETT

BANTAM PRESS

LONDON • TORONTO • SYDNEY • AUCKLAND • JOHANNESBURG

TRANSWORLD PUBLISHERS
61–63 Uxbridge Road, London W5 5SA
A Random House Group Company
www.rbooks.co.uk

First published in Great Britain
in 2009 by Bantam Press
an imprint of Transworld Publishers

A CIP catalogue record for this book
is available from the British Library.

ISBNs 9780593055267 (hb)
9780593055274 (tpb)

Addresses for Random House Group Ltd companies outside the UK
can be found at: www.randomhouse.co.uk
The Random House Group Ltd Reg. No. 954009

The Random House Group Limited supports The Forest Stewardship
Council (FSC), the leading international forest-certification organization. All our
titles that are printed on Greenpeace-approved FSC-certified paper carry the FSC logo.
Our paper procurement policy can be found at www.rbooks.co.uk/environment

Typeset in 11.5/15pt Bembo by
Falcon Oast Graphic Art Ltd.
Printed and bound in Great Britain by
CPI Mackays, Chatham, ME5 8TD

For my parents,

Sheila and Frank Beckett

1

Skin.

The largest human organ, it is also the most overlooked. Accounting for an eighth of the entire body mass, on an average adult it covers an area of approximately two square metres. Structurally skin is a work of art, a nest of capillaries, glands and nerves that both regulates and protects. It is our sensory interface with the outside world, the barrier at which our individuality – our *self* – ends.

And even in death, something of that individuality remains.

When the body dies, the enzymes that life has held in check run amok. They devour cell walls, causing the liquid contents to escape. The fluid rises to the surface, gathering below the dermal layers and causing them to loosen. Skin and body, until now two integral parts of the whole, begin to separate. Blisters form. Whole swathes begin to slip, sloughing off the body like an unwanted coat on a summer's day.

But, even dead and discarded, skin retains traces of its former self. Even now it can still have a story to tell, and secrets to keep.

Provided you know how to look.

★

Earl Bateman lay on his back, face turned to the sun. Overhead, birds wheeled in the blue Tennessee sky, cloudless but for the slowly dispersing vapour trail of a jet. Earl had always enjoyed the sun. Enjoyed the sting of it on his skin after a long day's fishing, enjoyed the way its brightness lent a new look to whatever it touched. There was no shortage of sun in Tennessee, but Earl came originally from Chicago, and the cold winters there had left a permanent chill in his bones.

When he'd moved to Memphis back in the seventies, he'd found the swampy humidity far more to his liking than the windy streets of his home city. Of course, as a dentist in a small practice, with a young wife and two small children to keep, he didn't spend as much time out in it as he might have liked. But it was there, all the same. He even liked the sweltering heat of Tennessean summers, when the breeze would feel like a hot flannel, and the evenings were spent in the airless swelter of the cramped apartment he and Kate shared with the boys.

Things had changed, since then. The dental practice had flourished, and the apartment had long since given way to bigger and better things. Two years before, he and Kate had moved into a new five-bedroomed house in a good neighbourhood, with a wide, rich green lawn where the growing brood of grandchildren could safely play, and the early morning sunshine would shatter into miniature rainbows in the fine spray from the water-sprinkler.

It had been on the lawn, sweating and cursing as he'd struggled to saw off a dead branch from the big old laburnum, that he'd had the heart attack. He'd left the saw still trapped in the tree limb and managed to take a few faltering steps towards the house before the pain had felled him.

In the ambulance, with an oxygen mask strapped over his face, he had held tightly on to Kate's hand and tried to smile to reassure her. At the hospital there had been the usual urgent ballet of medical staff, the frantic unsheathing of needles and beeping of machines. It had been a relief when they'd eventually fallen silent. A short time later, after the necessary forms had been signed, the inevitable bureaucracy that accompanies each of us from birth, Earl had been released.

Now he was stretched out in the spring sun. He was naked, lying on a low wooden frame that was raised off the carpet of meadow grass and leaves. He'd been here for over a week, long enough for the flesh to have melted away, exposing bone and cartilage under the mummified skin. Wisps of hair still clung to the back of his skull, from which empty eye sockets gazed at the cerulean blue sky.

I finished taking measurements and stepped out of the wire mesh cage that protected the dentist's body from birds and rodents. I wiped the sweat from my forehead. It was late afternoon and hot, despite the early season. Spring was taking its time this year, the buds swollen and heavy. In a week or two's time the display would be spectacular, but for now the birch and maples of the Tennessee woodland still hugged their new growth to them, as though reluctant to let go.

The hillside I was on was unremarkable enough. Scenic almost, though less dramatic than the imposing ridges of the Smoky Mountains that rose up in the distance. But it was an altogether different aspect of nature that struck everyone who visited here. Human bodies, in various stages of decay, lay all around. In the undergrowth, out in the full sun and lying in the shade; the more recent still bloated with decompositional gases, the older ones desiccated to leather. Some were hidden from view, buried underground or in car boots. Others, like the one I'd been weighing, were covered by mesh or chain-link screens, laid out like exhibits in some grisly art installation. Except that the purpose of this place was far more serious. And far less public.

I put my equipment and notepad back into my bag, flexing my hand to work the stiffness from it. A thin white line ran across my palm where the flesh had been laid open to the bone, cleanly bisecting the lifeline. Appropriately enough, given how the knife that had almost ended my life the previous year had also changed it.

I lifted the bag on to my shoulder and straightened. There was only the faintest of twinges from my stomach as I took the weight. The scar underneath my ribs was fully healed, and in another few weeks I'd be able to stop taking the antibiotics I'd been on constantly for the past nine months. I'd remain prone to infection for the rest of my life, but I counted myself lucky only to have lost a section of intestine along with my spleen.

It was what else I'd lost that I was finding harder to come to terms with.

Leaving the dentist to his slow decay, I skirted a body that lay partially hidden by shrubs, this one darkened and swollen, and followed the narrow dirt trail that meandered down through the trees. A young black woman in grey surgical smock and trousers was crouching by a half-hidden cadaver that was resting in the shade of a fallen tree trunk. She was using tweezers to pick squirming larvae from it, dropping each one into a separate screw-top jar.

'Hi, Alana,' I said.

She looked up and gave me a smile, tweezers poised. 'Hey, David.'

'Is Tom around?'

'Last I saw him he was down by the pads. And watch where you step,' she called after me. 'There's a district attorney in the grass down there.'

I raised my hand in acknowledgement as I carried on down the trail. It ran parallel to a high, chain-link fence that surrounded the two acres of woodland. The chain-link was topped with razor wire and screened by a second fence, this one made from timber. A large gate was the only way in or out, on which was hung a painted sign. In plain black letters were the words *Anthropology Research Facility*, but it was better known by another, less formal name.

Most people just called it the Body Farm.

The week before, I'd stood in the tiled hallway of my London flat, packed bags at my feet. A sweet chorus of birdsong sounded from the pale spring dawn outside. I ran through my mental list of things I needed to check, knowing I'd done everything already. Windows locked, post put on hold, boiler switched off. I felt edgy and ill at ease. I was no stranger to travelling, but this was different.

This trip there wouldn't be anyone waiting for me when I came back.

The taxi was late, but I had plenty of time to catch my flight. Still I found myself restlessly checking my watch. A few feet from where I stood, the black and white Victorian floor tiles caught my eye. I looked away, but not before the Harlequin pattern prompted the usual connection in my memory. The blood had long since been washed off the area next to the front door, just as it had from the wall above it. The entire hallway had been painted while I'd still been in hospital. There was no physical reminder of what had taken place here the previous year.

But all at once I felt claustrophobic. I carried my bags outside, careful not to put too much strain on my stomach. The taxi pulled up as I closed the front door. It shut behind me with a solid *thunk* that had a sound of finality about it. I turned away without a backward glance and walked to where the taxi was chugging out its fug of diesel fumes.

I took the cab only as far as the nearest tube station and caught the Piccadilly line to Heathrow. It was too early for the morning rush, but there were still people in the carriage, avoiding looking at each other with the instinctive indifference of the Londoner.

I'd be glad to leave, I thought, fervently. This was the second time in my life I'd felt the need to get away from London. Unlike the first, when I'd fled with my life in tatters after the death of my wife and daughter, I knew I'd be coming back. But I needed to escape for a while, to put some distance between myself and recent events. Besides which, I'd not worked in months. I hoped this trip would be a way of easing me back into things again.

And of finding out if I was still up to the job.

11

There was no better place to find out. Until recently, the facility in Tennessee had been unique, the only outdoor field laboratory in the world where forensic anthropologists used real human cadavers to study decomposition, recording the essential clues that might point to when and how death had occurred. A similar facility had now been set up in North Carolina, and also in Texas, once local concerns about vultures had been overcome. I'd even heard talk about one in India.

But it didn't matter how many there might be: in most people's minds the research facility in Tennessee was still *the* Body Farm. It was in Knoxville, part of the University of Tennessee's Forensic Anthropology Center, and I'd been lucky enough to train there early in my career. But it had been years since my last visit. Too long, as Tom Lieberman, its director and my old teacher, had told me.

As I sat in the departure lounge at Heathrow, watching the slow and silent dance of aircraft through the plate glass window, I wondered what it would be like going back. During the months of painful recovery after I came out of hospital — and the even more painful aftermath — the promise of the month-long trip had been something to work towards, a badly needed fresh start.

Now I was actually on my way, for the first time I wondered if I hadn't invested too much hope in it.

There was a two-hour stopover in Chicago before I caught my connecting flight, and the tail end of a storm was still grumbling as the plane landed in Knoxville. But it quickly cleared, and by the time I'd collected my baggage the sun was starting to break through. I breathed deeply as I left the airport terminal to collect my hire car, enjoying the unfamiliar humidity in the air. The roads steamed, giving off the peppery tang of wet tarmac. Against the slowly receding blue-black of the thunderheads, the rainfall gave the greens of the lush countryside around the highway an almost dazzling vibrancy.

I'd felt my spirits lift as I neared the city. *This is going to work.*

Now, barely a week later, I was no longer so sure. I followed the trail as it skirted a clearing in which stood a tall wooden tripod that resembled a bare tepee frame. A body lay on a platform beneath it, waiting to be hoisted and weighed. Leaving the trail – and remembering Alana's warning – I crossed the clearing to where several rectangular pads of concrete were set into the soil, starkly geometric in the woodland setting. Human remains were entombed in them, part of an experiment to see how effective ground-penetrating radar was in body location.

A tall, gangly figure in chinos and a floppy bush hat knelt a few yards away, scowling as he examined a gauge on a length of pipe protruding from the ground.

'How's it going?' I asked.

He didn't look up, peering through his wire-framed glasses as he gently nudged the gauge with a finger. 'You'd think it'd be easy to catch a smell this strong, wouldn't you?' he said by way of answer.

The flattened vowels betrayed his East Coast roots rather than the curling southern drawl of Tennessee. For as long as I'd known him, Tom Lieberman had been searching for his own Holy Grail, analysing the gases produced by decomposition molecule by molecule to identify the odour of decay. Anyone who'd ever had a mouse die under their floorboards could testify it existed, and it continued to exist long after human senses failed to detect it. Dogs could be trained to sniff out a cadaver years after it had been buried. Tom theorized that it should be possible to develop a sensor that would do much the same thing, making body location and recovery immeasurably easier. But, as with anything else, theory and practice were two very different things.

With a grunt that could have been either frustration or satisfaction he stood up. 'OK, I'm done,' he said, wincing as his knee joints cracked.

'I'm heading over to the cafeteria for some lunch. Are you coming?'

13

He gave a wistful smile as he packed away his equipment. 'Not today. Mary's packed sandwiches. Chicken and beansprouts, or something else disgustingly healthy. And before I forget, you're invited over for dinner this weekend. She seems to have got it into her head that you need a proper meal.' He pulled a face. 'You she wants to feed up; me, I just get rabbit food. Where's the justice in that?'

I smiled. Tom's wife was a great cook, and he knew it. 'Tell her I'd love to come. Do you want a hand with your gear?' I offered, as he hoisted his canvas bag on to his shoulder.

'No, it's OK.'

I knew he didn't want me to exert myself. But even though we walked slowly back to the gate I could see that the effort left him breathless. When I'd first met Tom he'd already been well into his fifties, happy to give encouragement to a fledgling British forensic anthropologist. That was longer ago than I cared to remember, and the intervening years had left their mark. We expect people to remain as we remember them, but of course they never do. Still, I'd been shocked at how changed Tom was when I saw him again.

He hadn't formally announced when he was stepping down as director of the Forensic Anthropology Center, but everyone knew it was likely to be before the end of the year. The local newspaper had run a feature on him two weeks earlier that had read more like a testimonial than an interview. He still looked like the basketball player he'd once been, but encroaching age had lent a gauntness to his already lean frame. There was a hollowness to his cheeks that, with the receding hairline, gave him an air that was both ascetic and worryingly frail.

But the twinkle in his eyes remained unchanged, as did his humour and a faith in human nature that was undimmed despite a career spent trawling through its darker side. *And you're not exactly unscathed yourself*, I reflected, remembering the ugly striation of flesh under my shirt.

Tom's station wagon was in the car park adjacent to the facility. We paused at the gate, pulling off the protective gloves and overshoes we'd been wearing before going out. With the barrier pulled shut behind us, there was nothing to suggest what lay on the other side. The trees behind the fence looked mundane and innocuous as they rustled in the warm breeze, bare branches shading green with new life.

Once we were in the car park I took my mobile from my pocket and switched it back on. Although there were no rules against it, I felt uncomfortable disturbing the peace and quiet inside the facility with phone calls. Not that I was expecting any. The people who might have contacted me knew I was out of the country, and the person I most wanted to talk to wouldn't be calling.

I put the phone away as Tom opened the boot and slid his bag into the back. He pretended not to be breathing heavily, while I pretended not to notice.

'Give you a lift to the cafeteria?' he offered.

'No thanks, I'll walk. I need the exercise.'

'Admirable discipline. You put me to shame.' He broke off as his phone rang. He took it out and glanced at the display. 'Sorry, got to take this.'

Leaving him to answer it, I headed across the car park. Although the facility was on the University of Tennessee Medical Center campus, it was completely independent of it. Tucked away on the wooded outskirts, it inhabited a different world. The modern buildings and park-like green spaces of the busy hospital were bustling with patients, students and medical staff. A nurse was laughing with a young man in jeans on a bench; a mother was scolding a crying child, while a businessman held an animated discussion on a mobile phone. When I'd first come here I'd found the contrast between the hushed decay behind the gates and the bustling normality outside them hard to take. Now I barely noticed it.

We can grow used to almost anything, given time.

I trotted up a flight of steps and set off along the path that led to the cafeteria, noting with satisfaction that I was breathing barely harder than usual. I'd not gone far when I heard footsteps hurrying behind me.

'David, wait up!'

I turned. A man about my own age and height was hurrying along the path. Paul Avery was one of the center's rising stars, already widely tipped as Tom's natural successor. A specialist in human skeletal biology, his knowledge was encyclopaedic, and the big hands and blunt fingers were as adept as any surgeon's.

'You going for lunch?' he asked, falling into step beside me. His curly hair was almost blue-black, and a shadow of stubble already darkened his chin. 'Mind if I join you?'

'Not at all. How's Sam?'

'She's good. Meeting Mary this morning to cruise around some of the baby stores. I'm expecting the credit card to take a serious hit.'

I smiled. I hadn't known Paul until this trip, but both he and his pregnant wife Sam had gone out of their way to make me welcome. She was nearly at full term with their first child, and while Paul did his best to appear blasé about it, Sam made no attempt to hide her excitement.

'Glad I saw you,' he went on. 'One of my PhD students has gotten engaged, so a few of us are going downtown tonight to celebrate. It'll be pretty relaxed, just dinner and a few drinks. Why don't you come along?'

I hesitated. I appreciated the offer, but the thought of going out with a group of strangers didn't appeal.

'Sam'll be going, and Alana, so you'll know some people there,' Paul added, seeing my reluctance. 'C'mon, it'll be fun.'

I couldn't think of a reason to say no. 'Well . . . OK, then. Thanks.'

'Great. I'll pick you up at your hotel at eight.'

A car horn honked from the road nearby. We looked back to see Tom's station wagon pulling up to the kerb. Winding down the window he beckoned us over.

'I just got a call from the Tennessee Bureau of Investigation. They've found a body in a mountain cabin out near Gatlinburg. Sounds interesting. If you're not busy, Paul, I thought you might want to come out with me and take a look?'

Paul shook his head. 'Sorry, I'm tied up all afternoon. Can't one of your graduate students help out?'

'They could, I suppose.' Tom turned to me, a sparkle of excitement in his eyes. Even before he spoke I knew what he was going to say. 'How about you, David? Care to do a little field work?'

2

The highway out of Knoxville streamed with slow-moving traffic. Even this early in the year it was warm enough to need the car's air conditioning. Tom had programmed the satnav to guide us when we reached the mountains, but for the moment we hardly needed it. He hummed quietly to himself as he drove, a sign I'd come to recognize as anticipation. For all the grim realism of the facility, the individuals who'd bequeathed their bodies there had all died natural deaths. This was different.

This was the real thing.

'So it looks like murder?' *Homicide*, I corrected myself. It was a safe bet, otherwise the Tennessee Bureau of Investigation wouldn't be involved. The TBI was a single-state version of the FBI, for whom Tom was a badge-carrying consultant. If the call had come from them rather than a local police department, then chances were that this was serious.

Tom kept his eyes on the road. 'Seems like it. I wasn't told much, but from the sound of things the body's in bad shape.'

I was starting to feel unaccountably nervous. 'Will there be any problem with me coming along?'

Tom looked surprised. 'Why should there be? I often take someone to help out.'

'I meant because I'm British.' I'd had to go through the usual red tape of visas and work permits in order to come out here, but I hadn't anticipated anything like this. I wasn't sure how welcome I'd be on an official investigation.

He shrugged. 'Can't see why that should be a problem. It's hardly national security, and I'll vouch for you if anyone asks. Or you could keep quiet and hope they don't notice your accent.'

Smiling, he reached to turn on the CD player. Tom used music the way other people smoked cigarettes or drank whisky, claiming it helped him to both clear his mind and focus his thoughts. His drug of choice was fifties and sixties jazz, and by now I'd heard the half-dozen albums he kept in the car often enough to recognize most of them.

He gave a little sigh, unconsciously settling back in the car seat as a track by Jimmy Smith pulsed from the speakers.

I watched the landscape of Tennessee slide past outside the car. The Smoky Mountains rose up ahead of us, shrouded in the blue-tinged mist for which they'd been named. Their forest-covered slopes stretched to the horizon, a rolling green ocean that was a stark contrast to the commercial bustle of the retail outlets around us. Garishly functional fast food outlets, bars and stores lined the high-way, the sky above them gridded with power lines and telegraph wires.

London and the UK seemed a long way away. Coming here had been a way to regain my edge and resolve some of the issues preying on my mind. I knew that there were some hard decisions to make when I got back. The temporary university contract I'd held in London had ended while I'd been convalescing, and although I'd been offered a permanent tenure, I'd received another offer from the forensic anthropology department of a top Scottish university. There had also been a tentative approach from the Forensic Search Advisory Group, a multi-disciplinary agency which helped the police locate bodies. It was all very flattering, and I should have been excited. But

19

I couldn't muster enthusiasm for any of it. I'd thought coming back here would change that.

So far it hadn't.

I sighed, rubbing my thumb across the scar on my palm without realizing it. Tom glanced across. 'You OK?'

I closed my hand on the scar. 'Fine.'

He accepted that without comment. 'Sandwiches are in my bag on the back seat. Might as well share them before we get there.' He gave a wry smile. 'Hope you like beansprouts.'

The country outside the car became more thickly wooded as we drew nearer the mountains. We drove through Pigeon Forge, a brash resort whose bars and restaurants chased along the roadside. One diner we passed was themed in a faux frontier style, right down to the plastic logs. A few miles further on we came to Gatlinburg, a tourist town whose carnival atmosphere seemed almost restrained in comparison. It had sprung up on the very edge of the mountains, and although its motels and shops clamoured for attention, they couldn't compete with the natural grandeur that rose up ahead.

Then we left it behind and entered another world. Steep, densely forested slopes closed in around us, plunging us into shadow as the road wound through them. Part of the huge Appalachian Mountains chain, the Smokies covered eight hundred square miles and spanned the border between Tennessee and North Carolina. They'd been declared a National Park, although looking out of the car window I thought that nature was blithely unaware of such distinctions. This was a wilderness that man had even now barely scratched. Coming from a crowded island like the UK, it was impossible not to be humbled by their sheer scale.

There was less traffic now. In a few weeks it would be much busier, but this was still spring and there were hardly any other cars to be seen. After a few more miles Tom turned off on to a gravelled side road.

'Shouldn't be much further now.' He checked the satnav display mounted on the dashboard, then peered up ahead. 'Ah, here we are.'

There was a sign saying *Schroeder Cabins, Nos 5–13* by a narrow track. Tom turned off on to it, the automatic transmission complaining slightly as it compensated for the gradient. Spaced well out from each other, I could make out the low-pitched roofs of cabins set back amongst the trees.

Police cars and unmarked vehicles I took to belong to the TBI lined both sides of the track ahead of us. As we approached, a uniformed police officer strode to block our way, hand resting lightly on the gun holstered on his belt.

Tom stopped and wound down the window, but the officer didn't give him time to speak.

'Sir, you cain't come up here. Y'all have to back up and leave.'

The accent was pure deep south, his politeness like a weapon in itself, implacable and unyielding. Tom gave him an easy smile.

'That's all right. Can you tell Dan Gardner that Tom Lieberman's here?'

The uniformed officer moved away a few paces and spoke into his radio. Whatever he heard reassured him.

''Kay. Park up there with the rest of the vehicles.'

Tom did as he was told. The nervousness I'd been feeling had solidified into a definite unease as we parked. I told myself that a few butterflies were understandable; I was still rusty from my convalescence, and I hadn't banked on working on an actual murder investigation. But I knew that didn't really account for it, even so.

'You sure it's all right my being here?' I asked. 'I don't want to tread on anyone's toes.'

Tom didn't seem concerned. 'Don't worry. Anyone asks, you're with me.'

We climbed out of the car. After the city, the air smelled fresh and clean, rich with the outdoor scents of wild flowers and loam. Late afternoon sunlight dappled through the branches, picking out the coiled green buds like fat emeralds. This high up, and in the shade of the trees, it was quite cool, which made the appearance of the man walking towards us even stranger. He was wearing a suit and tie, but the jacket was slung over one arm, and his pale blue shirt was stained dark with perspiration. His face was flushed and red as he shook Tom's hand.

'Thanks for coming. Wasn't sure if you were still on vacation.'

'Not any more.' Tom and Mary had only returned from Florida the week before I'd arrived. He'd told me he'd never been so bored in his life. 'Dan, I'd like you to meet Dr David Hunter. He's visiting the facility. I said it'd be OK for him to come along.'

It wasn't quite phrased as a question. The man turned to me. I'd have put him just the far side of fifty, his weathered, careworn face lined with deep creases. The greying hair was cut short, with a side parting that might have been drawn with a ruler.

He extended his hand. His grip was tight enough to be a challenge, the skin of his palm dry and calloused.

'Dan Gardner, Assistant Special Agent in Charge. Pleased t'meet you.'

I guessed the title was the equivalent of Senior Investigating Officer in the UK. He spoke with the distinctive twang of Tennessee, but the easy-going manner was deceptive. His eyes were sharp and appraising. *Reserving judgement.*

'So, what have you got?' Tom asked, reaching in the back of the station wagon for his case.

'Here, let me,' I said, lifting it out for him. Scar or no, I was in better shape than Tom to carry it. For once he didn't argue.

The TBI agent started back up the trail into the trees. 'Body's in a rental cabin. Manager found it this morning.'

'Definitely homicide?'

'Oh, yeah.'

He didn't enlarge. Tom gave him a curious glance but didn't press. 'Any ID?'

'Got a man's wallet with credit cards and a driver's licence, but we can't say for sure if they're the victim's. Body's too far gone for the photograph to be any use.'

'Any idea how long it might have been here?' I asked without thinking.

Gardner frowned, and I reminded myself I was only here to help Tom. 'I was kind of hoping you'd be able to tell us that,' the TBI agent answered, though to Tom rather than me. 'The pathologist's still here, but he can't tell us much.'

'Who's the pathologist? Scott?' Tom asked.

'No, Hicks.'

'Ah.'

There was a wealth of meaning in the way Tom said it, none of it complimentary. But right then I was more concerned with the way he was starting to labour a little on the uphill trail.

'Just a second,' I said. I set down his case and pretended to fasten my boot. Gardner looked irritated, but Tom drew in relieved breaths, making a show of wiping his glasses. He looked pointedly at the way the agent's shirt was darkened with perspiration.

'Hope you don't mind my asking, Dan, but are you all right? You seem . . . well, a little feverish.'

Gardner looked down at his damp shirt as though he'd only just noticed. 'Let's just say it's kinda hot in there. You'll see.'

We set off again. The trail levelled out as the woods parted to reveal a small, grassy clearing, paved with a gravel path clogged with weeds. Other paths forked off from it, all of them running to cabins barely visible amongst the trees. The one we were heading for was at the furthermost edge of the clearing, well away from the others. It was small, the outside clad in weather-faded timber. Bright yellow tape declaring POLICE LINE, DO NOT CROSS in bold black capitals had been strung across the path leading to its door, and there was the usual bustle of activity around it.

This was the first crime scene I'd attended in the US. In most regards it was the same as I was used to, but the subtle differences gave it an unreal quality. A group of TBI forensic agents in white overalls were standing by the cabin, their faces flushed and sweating as they drank thirstily from bottles of water. Gardner led us to where a young woman in a smart business suit was talking with an overweight man whose bald head shone like a polished egg. He was completely hairless, without even eyebrows or eyelashes. It gave him a look that was both newborn and slightly reptilian.

He turned as we approached, thin mouth splitting in a smile when he saw Tom. But it was a humourless one.

'Wondered when you'd show up, Lieberman.'

'Just as soon as I got the call, Donald,' Tom said.

'Surprised you needed one. Y'all could smell this one all the way to Knoxville.'

He chuckled, unperturbed that no one else seemed to find the joke funny. I guessed that this was Hicks, the pathologist Gardner had mentioned. The young woman he'd been talking to was slim, with the compact athleticism of a gymnast. She held herself with an almost military bearing, a look emphasized by the navy blue jacket and skirt and short-cropped dark hair. She wore no make-up, but didn't need it. Only her mouth let down the clinical appearance; full and curving, the lips hinted at a sensuality the rest of her seemed at pains to deny.

Her grey eyes settled on me briefly, expressionless but coldly assessing. Against the lightly tanned skin of her face, the whites seemed to shine with health.

Gardner made quick introductions. 'Tom, this is Diane Jacobsen. She's just joined the Field Investigations Unit. This is her first homicide, and I've been giving you and the facility a big boost, so don't let me down.'

She extended her hand, apparently unmoved by Gardner's attempt at humour. Tom's warm smile was met with the barest one of her own. I wasn't sure if the reserve was natural or if she was just trying too hard to be professional.

Hicks's mouth twitched with annoyance as he watched Tom. He

realized I was looking at him, and jerked his chin irritably in my direction.

'Who's this?'

He spoke as though I wasn't there. 'I'm David Hunter,' I said, even though the question hadn't been addressed to me. Somehow I knew there was no point in offering my hand.

'David's temporarily working with us out at the facility. He's kindly agreed to help me,' Tom said. 'Working with' was overstating it, but I wasn't going to quibble over the white lie.

'He's British?' Hicks exclaimed, picking up on my accent. I could feel my face burning as the young woman's cool stare settled on me again. 'You're letting tourists here now, Gardner?'

I'd known my presence might raise a few hackles, just as a stranger's would in a UK inquiry, but his attitude irked me all the same. Reminding myself I was Tom's guest, I bit back my response. Gardner himself looked far from happy as Tom cut in.

'Dr Hunter's here on my invitation. He's one of the top forensic anthropologists in the UK.'

Hicks gave an incredulous snort. 'You mean we don't have enough of our own?'

'I mean I value his expertise,' Tom said easily. 'Now, if we're done here, I'd like to make a start.'

Hicks shrugged with exaggerated politeness. 'Go ahead. Believe me, you're welcome to this one.'

He stalked off back towards the parked cars. Leaving the two TBI agents outside the cabin, Tom and I headed for a trestle table where boxes of disposable overalls, gloves, boots and masks had been set. I waited until we were out of earshot.

'Look, Tom, perhaps this isn't such a good idea. I'll wait in the car.'

He smiled. 'Don't mind Hicks. He works out of the morgue at UT Medical Center, so we cross paths occasionally. He hates having to defer to us in situations like this. Partly professional jealousy, but mainly because the man's an asshole.'

25

I knew he was trying to put me at ease, but I still felt uncomfortable. I was used to being at crime scenes, but I was acutely aware that I didn't belong at this one.

'I don't know . . .' I began.

'It isn't a problem, David. You'll be doing me a favour. Really.'

I let it go, but my doubts remained. I knew I should be grateful to Tom, that few British forensic experts ever get the opportunity to work a crime scene in the States. But for some reason I felt more nervous than ever. I couldn't even blame Hicks's hostility; I'd put up with a lot worse in my time. No, this was about me. At some point in the last few months I seemed to have lost my confidence along with everything else.

Come on, get a grip. You can't let Tom down.

Gardner came over to the trestle table as we were ripping open the plastic bags of overalls.

'You might want to strip down to your shorts under those. Pretty hot in there.'

Tom gave a snort. 'I haven't undressed in public since I was at school. I don't aim to start now.'

Gardner swatted at an insect buzzing round his face. 'Don't say I didn't warn you.'

I didn't share Tom's modesty, but I followed his example all the same. I felt enough out of place as it was, without stripping down to my boxers in front of everyone. Besides, it was only spring, and the sun was already starting to go down. How hot could it be in the cabin?

Gardner rummaged amongst the boxes until he found a jar of menthol rub. He smeared a thick dab under his nose, then offered it to Tom.

'You'll need this.'

Tom declined. 'No thanks. My sense of smell isn't what it used to be.'

Gardner silently held out the jar to me. Normally I didn't use it either. Like Tom I was no stranger to the odour of decomposition, and after spending the past week at the facility I'd become well and truly acclimatized to it. But I still accepted the jar, wiping the scented Vaseline on my top lip. My eyes instantly watered from the pungent vapour. I took a deep breath, trying to still my jangling nerves. *What the hell's wrong with you? You're acting like this is your first time.*

The sun was warm on my back as I waited for Tom to get ready. Low and dazzling, it brushed the tops of the trees as it made its slow descent into evening. It would come up again in the morning no matter what happened here, I reminded myself.

Tom finished zipping up his overalls and gave a cheery smile. 'Let's see what we've got.'

Pulling on our latex gloves, we walked up the overgrown path to the cabin.

3

The cabin door was closed. Gardner paused outside. He'd left his jacket with the boxes of overalls, and had put on a pair of plastic overshoes and gloves. Now he slipped on a white surgical mask. I saw him take a deep breath before he opened the door and we went inside.

I've seen human bodies in most states of death. I know how bad the different stages of putrefaction smell, can even differentiate between them. I've encountered bodies that have been burned to the bone, that have been reduced to soap-like slime after weeks under-water. None are pleasant, but it's an inevitable part of my work, and one I thought I was inured to.

But I'd never experienced anything like this. The stench was almost tangible. The nauseatingly sweet, bad-cheese stench of decomposing flesh seemed to have been distilled and concentrated, cutting through the menthol under my nose as though it wasn't there. The cabin was alive with flies, swirling excitedly around us, but they were almost incidental compared to the heat.

The inside of the cabin was like a sauna.

Tom grimaced. 'Good God . . .'

'Told you to wear shorts,' Gardner said.

The room was small and sparsely furnished. Several of the forensic team had broken off what they were doing to glance over as we'd gone in. Shuttered blinds had been pulled up to allow daylight in through the windows on either side of the door. The floor was black-painted boards covered with threadbare rugs. A pair of dusty antlers hung over a fireplace on one wall, while a stained sink, cooker and fridge stood against another. The rest of the furniture – TV, sofa and armchairs – had been roughly pushed to the sides, leaving the centre of the room clear, except for a small dining table.

The body was lying on it.

It was naked, spread-eagled on its back, arms and legs draped over the table edges. Swollen by gases, the torso resembled an overstuffed kitbag that had burst open. Maggots dripped from it to the floor, so many of them that they looked like boiling milk. An electric radiator stood next to the table, all three of its bars shimmering yellow. As I watched, a maggot dropped on to one of them and disappeared in a fat sizzle.

Completing the tableau was a hard-backed chair that had been positioned by the victim's head. It looked innocuous enough, until you thought to wonder why it was there.

Someone had wanted a good view of what they were doing.

None of us had gone any further than the doorway. Even Tom seemed taken aback.

'We left it like we found it,' Gardner said. 'Thought you'd want to record the temperature yourself.'

He went up a notch in my estimation. Temperature was an important factor in determining time since death, but not many investigating officers I'd come across would have thought of that. Still, on this occasion I almost wished he'd been less thorough. The combination of heat and stench was overpowering.

Tom nodded absently, his gaze already fixed on the body. 'Care to do the honours, David?'

I set his case down on a clear area of floorboards and opened it up.

Tom still had much of the same battered equipment he'd had since I'd known him, everything well worn and neatly ordered in its place. But while he might be a traditionalist at heart, he also recognized the benefits of new technology. He'd kept his old mercury thermometer, an elegant piece of engineering with its hand-blown glass and tooled steel, but alongside it was a new digital model. Taking it out, I switched it on and watched the numbers on its display quickly start to climb.

'How much longer will your people be?' Tom asked Gardner, glancing at the white-clad figures working in the room.

'A while yet. Too hot for them to stay long in here. I've had an agent pass out already.'

Tom was bending over the body, careful to avoid the dried blood on the floor. He adjusted his glasses to see better. 'Have we got a temperature yet, David?'

I checked the digital readout. I'd already started to sweat. 'Forty-three point five degrees.'

'So now can we turn off the goddamn fire?' one of the forensic team asked. He was a big man, with a barrel-like stomach that strained the front of his overalls. What was visible of his face under the surgical mask was red and sweating.

I glanced at Tom for confirmation. He gave a nod.

'Might as well open the windows too. Let's get some air in here.'

'Thank the sweet Lord for that,' the big man breathed as he went to unplug the fire. As its bars dimmed, he opened the windows as far as they would go. There were sighs and mutterings of relief as fresh air swept into the cabin.

I went to where Tom was staring down at the body with a look of abstract concentration.

Gardner hadn't been exaggerating; there was no question that this was a homicide. The victim's limbs had been pulled down on either side of the table and fastened to the wooden legs with parcel tape. The skin was drum-tight and the colour of old leather, although that was no indication of ethnicity. Pale skin darkens after death, while dark skin will often lighten, blurring colour and ancestry. What was more significant were the gaping slits that were evident. It's natural for the skin to split apart as the body decomposes and becomes bloated by gases. But there was nothing natural about this. Dried blood caked the table around the body and blackened the rug below it. That had to have come from an open wound, or possibly more than one, which suggested that at least some of the damage to the epidermis had been inflicted while the victim was still alive. It might also explain the numbers of blowfly larvae, as the flies would have laid their eggs in any opening they could find.

Even so, I couldn't recall ever seeing so many maggots in a single body before. Up close, the ammoniac stink was overpowering. They had colonized the eyes, nose, mouth and genitals, obliterating whatever sex the victim had been.

I found my eyes drawn to the way they seethed in the gaping slit in the stomach, causing the skin around it to move as though it were alive. My hand involuntarily went to the scar on my own.

'David? You OK?' Tom asked quietly.

I tore my gaze away. 'Fine,' I said, and began taking the specimen jars from the bag.

I could feel his eyes on me. But he let it pass, turning instead to Gardner. 'What do we know?'

'Not much.' Gardner's voice was muffled by his mask. 'Whoever did this was pretty methodical. No footprints in the blood, so the killer knew enough to mind where he put his feet. Cabin was rented out last Thursday to someone calling himself Terry Loomis. No description. Reservation and credit card payment were made by phone. Man's voice, local accent, and the guy asked for the key to be left under the mat by the cabin door. Said he'd be arriving late.'

'Convenient,' Tom said.

'Very. Don't seem too worried about paperwork here so long as they get paid. The cabin rental ended this morning, so when the key wasn't returned the manager came up to take a look and make sure nothing was missing. Place like this, you can see why he'd be worried,' he added, glancing round the threadbare cabin.

But Tom wasn't paying any attention. 'The cabin was only rented from last *Thursday*? You sure?'

'That's what the manager said. Date checks out with the register and the credit card receipts.'

Tom frowned. 'That can't be right. That's only five days ago.'

I'd been thinking the same thing. The decomposition was much too advanced for such a short period of time. The flesh was already displaying a cheesy consistency as it began to ferment and moulder, the leathery skin slipping off it like a wrinkled suit. The electric fire would have speeded things up to some extent, but that didn't explain the amount of larval activity. Even in the full heat and humidity of a Tennessean summer it would normally have taken nearer seven days to reach this stage.

'Were the doors and windows closed when he was found?' I asked Gardner without thinking. *So much for keeping quiet.*

He pursed his lips in displeasure, but still answered. 'Closed, locked and shuttered.'

I batted flies away from my face. You'd think I'd be used to them by now, but I'm not. 'A lot of insect activity for a closed room,' I said to Tom.

He nodded. Using tweezers, he carefully picked up a maggot from the body and held it up to the light to examine it. 'What do you make of this?'

I leaned closer to take a look. Flies have three larval stages, called instar, in which the larvae grow progressively larger.

'Third instar,' I said. That meant it had to be at least six days old, and possibly more.

Tom nodded, dropping the larva into a small jar of formaldehyde. 'And some of them have already started to pupate. That would make the time since death six or seven days.'

'But not five,' I said. My hand had strayed towards my stomach again. I took it away. *Come on, concentrate.* I made an effort to apply myself to what I was looking at. 'I suppose he could have been killed somewhere else and brought here post mortem.'

Tom hesitated. I saw two of the white-suited figures exchange a glance, and immediately realized my mistake. I felt my face burn. *Of all the stupid . . .*

'No need to tape the arms and legs to the table if the victim was already dead,' the big crime scene officer said, looking at me oddly.

'Maybe corpses in England are livelier than over here,' Gardner said, deadpan.

There was a ripple of laughter. I felt my face sting, but there was nothing I could say to make it any better. *Idiot. What's wrong with you?*

Tom fastened the lid back on to the killing jar, his face studiedly impassive. 'Think this Loomis is the victim or the killer?' he asked Gardner.

'Well, it was Loomis's driver's licence and credit cards that were in the wallet we found. Along with over sixty dollars in cash. We ran a check: thirty-six years old, white, employed as an insurance clerk in Knoxville. Unmarried, lives alone, and hasn't been in to work for several days.'

The cabin door opened and Jacobsen entered. Like Gardner she was wearing overshoes and gloves, but she managed to make even those look almost elegant. She wasn't wearing a mask, and her face was pale as she went to stand by the older agent.

'So, unless the killer booked the place in his own name and considerately left his ID behind, the likelihood is that this is either Loomis, or some other male we don't know about,' Tom said.

'That's about it,' Gardner said. He broke off as another agent appeared in the doorway.

'Sir, there's someone asking to see you.'

'I'll be right back,' Gardner said to Tom, and went outside.

Jacobsen remained in the cabin. Her face was still pale, but she folded her arms tightly in front of her as though restraining any weakness.

'How d'you know it's male?' she asked. Her eyes flicked automatically to the seething activity around the corpse's groin, but she quickly averted them again. 'I can't see anything to say either way.'

Her accent wasn't as strong as some I'd heard, but it was pronounced enough to mark her as local. I looked at Tom, but he was engrossed with the corpse. Or at least pretending to be.

'Well, apart from the size—' I began.

'Not all women are small.'

'No, but not many are as tall as this. And even a big woman would have a more delicate bone structure, especially the cranium. That's—'

'I know what a cranium is.'

God, but she was spiky. 'I was about to say that's usually a good indication of gender,' I finished.

Her chin came up, stubbornly, but she made no other comment. Tom straightened from where he'd been examining the gaping mouth.

'David, take a look at this.'

He moved aside as I went over. Much of the soft tissue had gone from the face; eyes and nasal cavity were heaving with maggots. The teeth were almost fully exposed, and where the gums had been the yellow-white of the dentine had a definite reddish hue.

'Pink teeth,' I commented.

'Ever come across them before?' Tom asked.

'Once or twice.' But not often. And not in a situation like this.

Jacobsen had been listening. 'Pink teeth?'

'It's caused by haemoglobin from the blood being forced into the dentine,' I told her. 'Gives the teeth a pinkish look under the enamel. You sometimes find it in drowning victims who've been in the water for some time, because they tend to float head down.'

34

'Somehow I don't think we're dealing with a drowning here,' Gardner said, clumping back into the cabin.

He had another man with him. The newcomer also wore overshoes and gloves but didn't strike me as either a police officer or a TBI agent. He was in his mid-forties, not plump exactly, but with a sleek, well-fed look about him. He wore chinos and a lightweight suede jacket over a pale blue shirt, and the well-fleshed cheeks were covered with a stubble that stopped just short of being a beard.

But the apparently casual appearance was a little too contrived, as though he'd styled himself on the chiselled models from magazine advertisements. The clothes were too well cut and expensive, the shirt open by one button too many. And the stubble, like the hair, was slightly too uniform to be anything other than carefully groomed.

He exuded self-assurance as he walked into the cabin. His half-smile never wavered as he took in the body tied to the table.

Gardner had dispensed with his mask, perhaps out of deference to the newcomer, who wasn't wearing one either. 'Professor Irving, I don't think you've met Tom Lieberman, have you?'

The newcomer turned his smile on to Tom. 'No, I'm afraid our paths haven't crossed. You'll have to excuse me if I don't shake hands,' he said, theatrically showing us his gloves.

'Professor Irving's a criminal personality profiler who's worked with the TBI on several investigations,' Gardner explained. 'We wanted to get a psychological perspective on this.'

Irving gave a self-deprecating grin. 'Actually, I prefer to call myself a "behaviouralist". But I'm not going to quibble about titles.'

You just have done. I told myself not to take my mood out on him.

Tom's smile was blandness itself, but I thought I detected a coolness about it. 'Pleased to meet you, Professor Irving. This is my friend and colleague, Dr Hunter,' he added, making up for Gardner's omission.

The nod Irving sent my way was polite enough, but it was obvious

I didn't register on his radar. His attention was already moving to Jacobsen, his smile widening.

'I don't think I caught your name?'

'Diane Jacobsen.' She seemed almost flustered, the cool she'd displayed so far in danger of slipping as she stepped forward. 'It's a pleasure to meet you, Professor Irving. I've read a lot of your work.'

Irving's smile broadened even further. I couldn't help but notice how unnaturally white and even his teeth were.

'I trust it met with your approval. And, please, call me Alex.'

'Diane majored in psychology before she joined the TBI,' Gardner put in.

The profiler's eyebrows rose. 'Really? Then I'll have to be extra careful not to slip up.' He didn't actually pat her on the head, but he might as well have. An expression of distaste replaced his smile as he considered the body. 'Seen better days, hasn't it? Can I have a little more of that menthol, please?'

The request wasn't addressed to anyone in particular. After a moment one of the forensic team grudgingly went out to get it. Steepling his fingers, Irving listened without comment as Gardner briefed him. When the agent returned, the profiler accepted the menthol without acknowledgement, dabbing a neat smear on his top lip before holding out the jar for her to take.

She looked down at the proffered jar before taking it. 'Any time.'

If Irving was aware of the sarcasm he gave no sign. Tom shot me an amused look as he took another specimen jar from the bag and turned back to the body.

'I'd rather you wait till I'm done, please.'

Irving spoke without looking at him, as though taking for granted that everyone there would naturally defer to his wishes. I saw annoyance flash in Tom's eyes, and for a moment I thought he was going to respond. But before he could a sudden spasm crossed his face. It was gone so quickly I might have imagined it, except for the pallor it left behind.

'Think I'll get some fresh air. Too damn hot in here.'

He looked unsteady as he headed for the door. I started to go after him but he stopped me with a shake of his head.

'No need for you to come. You can start taking photographs once Professor Irving's finished. I'm just going to get some water.'

'There's iced bottles in a cooler by the tables,' Gardner told him.

I felt concerned as I watched him go, but it was clear Tom didn't want to make a fuss. No one else seemed to have noticed anything was wrong. He'd been facing away from everyone except Irving and me, and the profiler was oblivious anyway. He stood with his hand on his chin as Gardner resumed his briefing, staring intently at the dead man on the table. When the TBI agent had finished he didn't move or speak, his pose one of deep contemplation. *Pose being the operative word.* I told myself not to be uncharitable.

'You realize it's a serial, of course?' he said, stirring at last.

Gardner looked pained. 'We don't know that for sure.'

Irving's smile was condescending. 'Oh, I think we do. Look at the way the body's been *arranged*. It's been put on display for us to find. Stripped, bound, and in all probability tortured. And then left face up. There's no sign of any shame or regret, no attempt to cover the victim's eyes or turn him face down. This whole thing shouts of calculation and enjoyment. He was *pleased* with what he'd done, that's why he wanted you to see it.'

Gardner accepted the news with resignation. He must have known as much himself. 'So the killer's male?'

'Of course he is.' Irving chuckled as though Gardner had made a joke. 'Apart from everything else, the victim was obviously a powerful man. You think a woman's capable of doing this?'

You'd be surprised what some women are capable of. I could feel my scar starting to itch.

'We're looking at a huge, *huge* amount of arrogance here,' Irving went on. 'The killer must have known the body would be found when the rental period was up. My God, he even left the wallet so you could ID the victim. No, this was no one-off. Our boy's just getting started.'

The prospect seemed to please him.

'The wallet might not be the victim's,' Gardner said half-heartedly.

'I disagree. The killer's been far too deliberate to have left his own behind. I'd lay odds that he even made the reservation for the cabin himself. He didn't just happen along and decide to kill whoever was renting it. This was too well planned, too well orchestrated for that. No, he made the booking in the victim's name, then brought him out here. Somewhere nice and isolated, no doubt scouted in advance, where he could torture him at leisure.'

'How can you be sure the victim was tortured?' Jacobsen said. It was the first time she'd spoken since Irving had patronized her.

The profiler seemed to be enjoying himself. 'Why else tie him to the table? He wasn't just restrained, he was *staked out*. The killer wanted to take his time over this, to enjoy it. I don't suppose there's any way to check for semen deposits or evidence of sexual assault?'

It took me a moment to realize that this last question was aimed at me. 'Not when the body's this badly decomposed, no.'

'Pity.' He made it sound as though he'd missed a dinner party invitation. 'Still, from the amount of blood on the floor, it's obvious that the wounding was done while the victim was still alive. And I think the genital mutilation's highly significant.'

I spoke automatically. 'Not necessarily. Blowflies will lay their eggs around any body opening, including the groin. The insect activity doesn't mean there was a wound there. We'll need to carry out a full examination to determine that.'

'Really.' Irving's smile had set. 'But you'll allow that the blood came from *somewhere*? Or is the mess under the table just spilt coffee?'

'I was just pointing out that—' I began, but Irving was no longer listening. I clamped my mouth shut, angrily, as he turned to Gardner and Jacobsen.

'As I was saying, we've got a bound and naked victim who was tied down and in all probability mutilated. The question is whether the wounds were the result of post-coital rage, or frustrated sexual tension. In other words, were they inflicted *because* he got it up, or because he *didn't*?'

His words were met by silence. Even the forensic team had broken off to listen.

'You think the motivation's sexual?' Jacobsen asked, after a moment.

Irving feigned surprise. I felt my dislike of him edge up a little more.

'I'm sorry, I thought that would have been obvious from the fact the victim was left naked. That's why the wounding is important. We're dealing with someone who is either in denial about his sexuality, or who resents it and takes out his self-disgust on his victim. Either way, he isn't openly homosexual. He could be married, a pillar of society. Perhaps someone who likes to boast about his female conquests. This was done by someone who hates what he is, and who sublimated that self-loathing into aggression against his victim.'

Jacobsen's face was expressionless. 'I thought you said the killer was proud of what he'd done? That there was no sign of shame or regret?'

'Not over the actual killing, no. He's beating his chest here, trying to convince everyone – including himself – how big and tough he is. But the *reason* he did it, that's another matter. *That's* what he's ashamed of.'

'There could be other reasons why the victim's naked,' Jacobsen said. 'Could be a form of humiliation or another way to exercise control.'

'One way or another, control usually comes down to sex.' Irving smiled, but it was starting to look a little forced. 'Gay serial killers are rare, but they do exist. And from what I've seen I think that may well be what we've got here.'

Jacobsen wasn't about to back down. 'We don't know enough about the killer's motivation to—'

'Forgive me, but do you have much experience with serial killer investigations?' Irving's smile had frost on it.

'No, but—'

'Then perhaps you'd spare me the pop psychology.'

There wasn't even the pretence of a smile now. Jacobsen didn't react, but the twin patches of red on her cheeks betrayed her. I felt sympathy for her. Outspoken or not, she hadn't deserved that.

An awkward silence had descended. Gardner broke it. 'What about the victim? You think the killer might have known him?'

'Maybe, maybe not.' Irving seemed to have lost interest. He was tugging at the collar of his shirt, the rounded face flushed and beaded with sweat. The cabin had cooled since the window had been opened, but it was still stiflingly hot. 'I'm done here. I'll need copies of forensic reports and photographs, along with whatever inform-ation you have on the victim.'

He turned to Jacobsen with what I imagine he thought was an engaging grin. 'Hope you didn't mind our little difference of opinion. Perhaps we could discuss it at more length over a drink sometime.'

Jacobsen didn't answer, but the way she looked at him made me think he shouldn't build up his hopes. The profiler was wasting his time if he was trying to charm her.

The atmosphere in the small cabin became more relaxed once Irving had left. I went to get the camera from Tom's case. It was a cardinal rule to take our own photographs of the body, regardless of whatever crime scene ones there were. But before I could start a shout went up from one of the agents.

'Think I've got something.'

It was the big man who'd spoken. He was kneeling on the floor by the sofa, straining to reach underneath. He pulled out a small grey cylinder, holding it with surprising delicacy in his gloved fingers.

'What is it?' Gardner asked, going over.

'Looks like a film canister,' he said, breathless from the effort. 'For a thirty-five-millimetre camera. Must've rolled under there.'

I glanced at the camera I had in my hand. Digital, the same as most forensic investigators used nowadays.

'Does anyone still *use* film?' asked the female agent who'd fetched Irving the menthol.

'Only die-hards and purists,' the big man said. 'My cousin swears by it.'

'He into glamour photography like you, Jerry?' the woman asked, raising a laugh.

But Gardner's face didn't slip. 'Anything inside?'

The big agent peeled off the lid. 'Nope, only air. Wait a second, though . . .'

He held the shiny cylinder up to the light, squinting along its length.

'Well?' Gardner prompted.

I could see the agent called Jerry grin even though he was wearing a mask. He waggled the film container.

'Can't offer you any photographs. But will a nice fat fingerprint do instead?'

The sun was setting as Tom drove us back towards Knoxville. The road wound through the bottom of steep, tree-covered slopes that blocked out the last of the light, so that it was dark even though the sky above us was still blue. When Tom flicked on the headlights, night suddenly closed in around us.

'You're quiet,' he said after a while.

'Just thinking.'

'I kind of guessed that.'

I'd been relieved to see he looked much better when he'd returned to the cabin. The rest of the work had gone smoothly enough. We'd photographed and sketched the position of the body, then taken tissue samples. By analysing the amino and volatile fatty acids released as the cells broke down we'd be able to narrow the time since death to within twelve hours. At the moment everything pointed to the victim's being dead for at least six days, and very possibly seven. Yet according to Gardner the cabin had only been occupied for five. Something wasn't right, and although I might have lost confidence in my own abilities, I was certain of one thing.

Nature didn't lie.

I realized Tom was waiting for me to respond. 'I didn't exactly cover myself in glory back there, did I?'

'Don't be too hard on yourself. Everyone makes mistakes.'

'Not like that. It made me look like an amateur. I wasn't thinking.'

'C'mon, David, it wasn't such a big deal. Besides, you might still be right. There's something skewed about the time since death. Maybe the victim was already dead when he was taken to the cabin. The body could have been tied to the table to make it look like he'd been killed there.'

Much as I'd have liked to believe that, I couldn't see it. 'That would mean the entire crime scene was staged, including the blood on the floor. And anyone clever enough to make it as convincing as that would know it wouldn't fool us for long. So what would be the point?'

Tom had no answer to that. The road marched between silent walls of trees, their branches picked out starkly in the headlights.

'What did you make of Irving's theory?' he asked after a while.

'You mean this being the start of a serial spree, or that it was sexually motivated?'

'Both.'

'He could be right about it being a serial killer,' I said. Most murderers tried to conceal their crimes, hiding their victims' bodies rather than leaving them on display. This smacked of a very different sort of killer, with a very different agenda.

'And the rest?'

'I don't know. I'm sure Irving's good at what he does, but . . .' I gave a shrug. 'Well, I thought he was too eager to jump to conclusions. It seemed to me like he was seeing what he wanted to rather than what was actually there.'

'People who don't understand what we do might think the same about us.'

'At least what we do is based on hard evidence. Irving seemed to me to be speculating an awful lot.'

'Are you saying you never listen to your instincts?'

'I might listen, but I wouldn't let them get in the way of the facts. Neither would you.'

He smiled. 'I seem to recall that we've had this discussion before. And no, of course I'm not saying we should rely on instinct too much. But used judiciously it's another tool at our disposal. The brain's a mysterious organ; sometimes it makes connections we're not consciously aware of. You've got good instincts, David. You should learn to trust them more.'

After my blunder in the cabin that was the last thing I wanted to do. But I wasn't going to let this turn into a discussion about me. 'Irving's whole approach was subjective. He seemed too keen for the killer to be a repressed homosexual, something nice and sensational. I got the impression he was already planning his next paper.'

Tom gave a laugh. 'More likely his next book. He made the bestseller charts a couple of years ago, and since then he's been a head for hire for any TV company that'll pay his fees. The man's a shameless self-promoter, but in fairness he has had some good results.'

'And I bet they're the only ones anyone hears about.'

Tom's glasses caught the reflection from the headlights as he gave me a sideways glance. 'You sound very cynical these days.'

'I'm just tired. Don't pay any attention.'

Tom turned back to the road. I could almost feel the question coming. 'This is none of my business, but what happened with the girl you were seeing? Jenny, wasn't it? I haven't wanted to mention it before, but . . .'

'It's over.'

The words seemed to have an awful finality to them, one that still didn't seem to apply to me and Jenny.

'Because of what happened to you?'

'That was part of it.' That and other things. *Because you put your work first. Because you were nearly killed. Because she didn't want to sit at home any more, wondering if it was going to happen again.*

'I'm sorry,' Tom said.

I nodded, staring dead ahead. *So am I.*

The indicator clicked as he turned off on to another road. This one seemed even darker than the last.

'So how long have you had a heart problem?' I asked.

Tom said nothing for a second, then gave a snort. 'I keep forgetting about that damn medical background of yours.'

'What is it, angina?'

'So they say. But I'm fine, it's not serious.'

It had looked serious enough to me that afternoon. I thought about all the other times I'd seen him having to stop to catch his breath since I'd arrived. I should have realized sooner. If I hadn't been so wrapped up in my own problems perhaps I would.

'You should be taking it easy, not trekking up hillsides,' I told him.

'I'm not about to start babying myself,' he said irritably. 'I'm on medication, it's under control.'

I didn't believe him, but I knew when to back off. We drove in silence for a while, both of us aware of things left unsaid. The inside of the station wagon was lit up as another car came up behind us, its headlights dazzlingly bright.

'So how do you feel about lending me a hand with the examination tomorrow?' Tom asked.

The body was going to be taken to the morgue at UT Medical Center in Knoxville. As a visual ID was out of the question, trying to identify the body was a priority. The Forensic Anthropology Center had its own lab facilities – bizarrely based at Neyland sports stadium in Knoxville – but they were more often used for research rather than actual homicide investigations. The TBI also had its own facilities in Nashville, but the UTMC morgue was more convenient in this instance. Normally, I would have jumped at the opportunity to help Tom, but now I hesitated.

'I'm not sure I'm up to it.'

'Bullshit,' Tom said, uncharacteristically blunt. He gave a sigh. 'Look, David, you've had a tough time lately, I know that. But you came over here to get back on your feet, and I can't think of a better way to do it.'

'What about Gardner?' I hedged.

'Dan's a little prickly with people he doesn't know sometimes, but he appreciates talent as much as anyone. Besides, I don't have to ask his permission to get someone to help me. I'd normally use one of my students, but I'd rather have you there. Unless you don't *want* to work with me, of course.'

I didn't know what I wanted, but I could hardly turn him down. 'If you're sure, then thanks.'

Satisfied, he turned his attention back to the road ahead. Suddenly, the inside of the car was flooded with light as the car behind us closed the gap. Tom squinted as its headlights dazzled him in the rear-view mirror. They were only a few feet away, high and bright enough to suggest they belonged to either a pick-up or a small truck.

Tom clicked his tongue in annoyance. 'What the hell's this idiot doing?'

He slowed, pulling over to the side of the road to let the other car pass. But its headlights slowed as well, remaining right behind us.

'Fine, you've had your chance,' Tom muttered, speeding up again.

The headlights kept pace with us, staying just behind the station wagon. I twisted round, trying to see what was following us. But the glare rendered everything through the rear window invisible, prevented me from making anything out.

With a screech of rubber, the headlights abruptly swerved to the left. I caught a glimpse of a high-bodied pick-up, its windows black mirrors as it tore past with a throaty roar. The station wagon was rocked by its slipstream and then it was gone, its rear lights quickly disappearing into the darkness.

'Damn redneck,' Tom muttered.

He reached for the CD player, and the mellow tones of Chet Baker accompanied us back to civilization.

4

Tom dropped me off at the hospital where I'd left my car. We arranged to meet first thing next morning at the morgue, and after he'd gone I gratefully drove back to my hotel. All I wanted to do was have a shower, get something to eat and then try to sleep.

Which was pretty much what I'd done almost every night so far.

I was on my way up to my room before I remembered I'd agreed to go out that evening. I checked the time and saw I'd less than half an hour before Paul was due to pick me up.

I sank down on to the bed with a groan. I felt less like company than ever. I was out of the habit of socializing, and the last thing I was in the mood for was making polite conversation with strangers. I was tempted to call Paul and make some excuse, except I couldn't think of one. Besides, it would be churlish to turn down their hospitality.

Come on, Hunter, make an effort. God forbid you should enjoy yourself. Reluctantly, I pushed myself off the bed. There was just enough time for a shower if I hurried, so I stripped off my clothes and stepped into the cubicle, turning the jet on full. The scar on my stomach looked alien and strange, as if it wasn't really a part of me. Even though the ugly line of pink flesh wasn't tender any more, I still didn't like touching it. In time I supposed I'd become used to its presence, but I wasn't yet.

I turned my face up to the stinging spray, taking deep breaths of the steam-filled air to dispel the sudden rush of memory. *The knife handle protruding from below my ribs, the hot, sticky feel of blood pooling around me on the black and white tiles . . .* I shook my head like a dog, trying to cast out the unwanted images. I'd been lucky. Grace Strachan was one of the most beautiful women I'd ever known. She was also the most dangerous, responsible for the deaths of at least half a dozen people. If Jenny hadn't found me in time I'd have added to that tally, and while I knew I should be grateful to be alive, I was finding it hard to put it behind me.

Especially since Grace was still out there.

The police had assured me that it was only a matter of time before she was found, that she was too unstable to remain free for long. But Grace had been a rich woman, consumed by a passion for vengeance that was as irrational as it was deadly. She wasn't going to give herself away that easily. Nor was I her only target. She'd already tried to kill a young mother and daughter once, and only been prevented at the cost of another life. Since Grace's attack on me, Ellen and Anna McLeod had been living under police protection and an assumed name. While they'd prove harder to track down than a forensic scientist who was listed in the phone book, the truth was that none of us would be safe until Grace was caught.

That wasn't an easy thing to live with. Not when I bore the scars to remind me how close she'd come already.

I turned up the shower as hot as I could stand it, letting the water scald away the dark thoughts. Dripping wet, I towelled myself dry until my skin was stinging, then dressed and hurried downstairs. The hot shower made me feel better, but I still felt little enthusiasm as I went down to the hotel foyer. Paul was already there, scribbling intently in a small notepad as he waited on a sofa.

'Sorry, have you been waiting long?' I asked.

He stood up, tucking the notepad into a back pocket. 'Only just got here. Sam's in the car.'

He'd parked across the street. A pretty woman in her early thirties was waiting in the passenger seat. She had long, very blond hair and turned to face me as I slid into the back, her hands resting on her swollen stomach.

'Hey, David, good to see you again.'

'You too,' I said, meaning it. There are some people you feel instantly at ease with, and Sam was one of them. We'd only met once, earlier that week, but it already seemed like I'd known her for years. 'How are you feeling?'

'Well, my back hurts, my feet ache, and you don't even want to know about the rest. But other than that I can't complain.' She smiled to show she didn't mean it. Sam was one of the lucky women who wear their pregnancy well. She fairly shone with health, and for all the discomfort it was obvious she was loving every moment.

'Junior's been playing up lately,' Paul said, pulling out into the traffic. 'I keep on telling Sam that's a sure sign it's a girl, but she won't listen.'

Neither of them had wanted to know the sex of the baby. Sam had told me it would have spoiled the surprise. 'Girls aren't that boisterous. It's a boy.'

'Case of beer says you're wrong.'

'A case of *beer*? That's the best you can do?' She appealed to me. 'David, what sort of bet is that for a pregnant woman?'

'Sounds pretty shrewd to me. He gets to drink it even if he loses.'

'Hey, you're supposed to be on my side,' Paul protested.

'He's got more sense,' Sam said, swatting him.

I began to unwind as I listened to their banter. It felt good to see their happiness, and if I felt a tug of envy it was only a small one. When Paul pulled up into a parking space I was disappointed the short journey was over.

We were in the Old City, the one-time industrial heart of

Knoxville. Some factories and warehouses still remained, but the area had undergone a genteel conversion, the industry giving way to bars, restaurants and apartments. Paul had parked a little way up the street from the steakhouse where everyone was meeting, an old brick building whose cavernous space was now filled with tables and live music. It was already busy, and we had to ease our way to a large group sitting by one of the windows. The half-empty beer glasses and laughter announced that they'd been there for some time, and for a second I faltered, wishing I'd not come.

Then space was found for me at the table, and it was too late. Introductions were made, but I forgot the names as soon as I heard them. Other than Paul and Sam, the only person I recognized was Alana, the forensic anthropologist who'd told me where to find Tom in the facility earlier. She was with a brawny man I guessed must be her husband, but the rest were either faculty members or students I didn't know.

'You've got to try the beer, David,' Paul said, leaning round Sam to see me. 'This place has its own microbrewery. It's fantastic.'

I'd hardly touched alcohol in months, but I felt I needed something now. The beer was a dark brew served cold, and tasted wonderful. I drank half of it almost straight off, and set the glass down with a sigh.

'You look like you needed that,' Alana said from across the table. 'One of those days, huh?'

'Something like that,' I agreed.

'Had a few of those myself.'

She raised her glass in an ironic toast. I took another drink of beer, feeling myself begin to relax. The atmosphere around the table was informal and friendly, and I slipped easily into the conversations going on around me. When the food arrived I tore into it. I'd ordered steak and a green salad, and I hadn't realized how hungry I was until then.

'Having fun?'

Sam was grinning at me over the top of her glass of mineral water. I nodded, working to swallow a mouthful of steak.

'Is it that obvious?'

'Uh-huh. First time I've seen you look relaxed. You should try it more often.'

I laughed. 'I'm not that bad, am I?'

'Oh, just wound a little tight.' Her smile was warm. 'I know you came here to get some things straightened out. But there's no law says you can't enjoy yourself from time to time. You're among friends, you know.'

I looked down, more affected than I wanted to admit. 'I know. Thanks.'

She shifted in her seat and winced, putting her hand to her stomach.

'Everything OK?' I asked.

She gave a pained smile. 'He's a little restless.'

'He?'

'He,' she said firmly, stealing a look across at Paul. 'Definitely he.'

The plates were cleared away, desserts and more drinks ordered. I had coffee, knowing if I had another beer I'd regret it in the morning. I leaned back in my chair, savouring the slight buzz of well-being.

And then my good mood crashed around me.

From nowhere I caught a waft of musk, lightly spiced and un-mistakable. A second later it had vanished, lost amongst the stronger odours of food and beer, but I knew I hadn't imagined it. Recognition ran through me like an electric shock. For an instant I was back on the tiled floor of my hallway, the metallic stink of blood blending with a more delicate, sensual scent.

Grace Strachan's perfume.

She's here. I bolted upright in my seat, frantically looking around. The restaurant was a confusion of sound and colour. I scanned the

51

faces, desperately searching for a telltale feature, some flaw in a disguise. *She must be here somewhere. Where is she?*

'Coffee?'

I stared blankly up at the waitress who'd appeared next to me. She was in her late teens, a little overweight. Her perfume cut through the cooking and bar-room smells: a cheap musk, heavy and cloying. Up close, it was nothing like the subtle perfume that Grace Strachan used.

Just similar enough to fool me for a second.

'You order coffee?' the waitress prompted, giving me a wary look.

'Sorry. Yes, thank you.'

She set it down and moved on. My arms and legs prickled, shivery with the aftermath of adrenalin. I realized my hand was clenched so tightly around its scar that it hurt. *Idiot. As if Grace could have followed you . . .* Awareness of how brittle my nerves were, even here, left a sour taste in my mouth. I tried to force myself to relax but my heart was still racing. All at once there didn't seem to be enough air in the room. The noise and smells were unbearable.

'David?' Sam was looking at me with concern. 'You've gone white as a sheet.'

'I'm just a little tired. I'm going to head on back.' I had to get outside. I started fumbling notes from my wallet, not seeing what they were.

'Wait, we'll drive you.'

'No!' I put my hand on her arm before she could turn to Paul. 'Please. I'll be fine, really.'

'You sure?'

I made myself smile. 'Certain.'

She wasn't convinced, but I was already pushing my chair back, dropping a handful of notes on to the table without knowing if it was enough or not. Paul and the others were still busy talking, but I didn't stop to see if anyone else noticed me leave. It was all I could do not to break into a run as I barged through the door into the street. I sucked in deep breaths of the cool spring air, but didn't stop even then. I kept walking, not knowing or caring where I was heading, wanting only to keep moving.

I stepped off the kerb and jumped back as a horn blared deafeningly to my left. I stumbled back on to the pavement as a trolley car rattled past inches in front of me, its windows bright splashes of light in the darkness. As soon as it had passed I hurried across the road, taking turnings at random. It had been years since I'd been to Knoxville, and I had no idea of where I was and even less of where I was going.

I didn't care.

It was only when I saw the stretch of blackness beyond the streetlights ahead of me that I finally slowed. I could feel the river even before I saw it, a moistness in the air that finally brought me back to myself. I was drenched in sweat as I leaned on the railings. The bridges that spanned the tree-lined banks were skeletal arches in the darkness, dotted with lights. Below them, the Tennessee river sedately idled past, just as it had for thousands of years. And probably would for thousands more.

What the hell's wrong with you? Running scared just because of a cheap perfume. But I felt too wrung out to be ashamed. Feeling as alone as I ever had in my life, I took my phone out and scrolled through my contacts. Jenny's name and number were highlighted on the illuminated display. I held my thumb poised over the dial key, badly wanting to talk to her again, to hear her voice. But it was the early hours of the morning back in the UK, and even if I called her, what would I say?

It had all been said already.

'Got the time?'

I gave a start as the voice came from beside me. I was in an area of darkness between streetlights, and all I could make out of the man was the red glow of his cigarette. Belatedly, I realized that the street was deserted. *Stupid. All this way just to get mugged.*

'Half past ten,' I told him, tensing for the attack that would come next.

But he only gave a nod of thanks and walked on, disappearing into the dark beyond the next streetlight. I shivered, and not only because of the damp chill coming from the river.

The welcoming yellow lights of a taxi were approaching on the lonely street. Flagging it down, I went back to my hotel.

The cat is your earliest memory.

There must be others before it, you know that. But none so vivid. None that you take out and replay time after time. So real that even now you can still feel the sun on the back of your head, see your shadow on the ground in front of you as you bend over.

The soil is soft and easy to turn. You use a piece of wood broken off the fence, a piece of white picket starting to soften and rot. It threatens to break again, but you don't have far to dig.

It isn't deep.

You smell it first. A cloying, sweet stink that's both familiar and like nothing you've smelled before. You stop for a while, sniffing the damp soil, nervous but more excited. You know you shouldn't be doing this, but the curiosity is too great. Even then you had questions; so many questions. But no answers.

The wood hits something almost as soon as you continue digging. A different texture in the soil. You begin to scrape away the final covering of earth, noticing that the smell has grown stronger. Finally, you can see it: a cardboard shoebox, its sides soaked and rotting.

The box starts to disintegrate when you try to lift it, wet and sagging from the weight inside. You quickly set it down again. Your fingers feel clumsy and strange as you take hold of the lid, your chest tight. You're scared, but excitement easily outweighs your fear.

Slowly, you remove the shoebox lid.

The cat is a dirty mound of ginger. Its half-closed eyes are pale and dull, like deflated balloons after a party. Insects are crawling in its fur, beetles scuttling from the daylight. You stare, rapt, as a fat worm coils and contracts, dripping from its ear. Taking the stick, you prod the cat. Nothing happens. You prod again, harder. Again, nothing. A word forms in your mind, one you've heard before, but never really comprehended until now.

Dead.

You remember the cat as it was. A fat, bad-tempered tom, a thing of spite and claws. Now it's . . . nothing. How can the living animal you remember have become this rotting clump of fur? The question fills your head, too huge for you to hold. You lean closer, as though if you look hard enough you'll find the answer . . .

. . . and suddenly you're jerked away. The neighbour's face is contorted with anger, but there's also something there you don't recognize. It's only years later that you identify it as disgust.

'What in God's name are you . . . ? Oh, you sick little bastard!'

There is more shouting, then and later, back at the house. You don't try to explain what you did, because you don't understand yourself. But neither the angry words nor the punishment wipe away the memory of what you saw. Or what you felt, and still feel even now, nestling in the pit of your stomach. An overwhelming sense of wonder, and of burning, insatiable curiosity.

You're five years old. And this is how it starts.

5

Everything seemed to slow down as the knife came towards me. I grabbed for it, but I was always going to be too late. The blade slid through my grip, slicing my palm and fingers to the bone. I could feel the hot wetness of blood smearing my hand as my legs gave way under me. It pooled on the black and white floor tiles as I slid down the wall, soaking the front of my shirt.

I looked down and saw the knife handle protruding obscenely from my stomach and opened my mouth to scream . . .

'No!'

I bolted upright, gasping. I could feel the blood on me, hot and wet. I thrashed off the sheets, frantically trying to see my stomach in the dim moonlight. But the skin was unmarked. There was no knife, no blood. Just a sheen of clammy sweat, and the angry welt of the scar just under my ribs.

Christ. I sagged with relief, recognizing my hotel room, seeing I was alone in it.

Just a dream.

My heart rate was starting to return to normal, my pulse ebbing in my ears. I swung my legs off the edge of the bed and shakily sat up. The clock on the bedside cabinet said five thirty. The alarm was set for an hour's time, but it wasn't worth trying to sleep again, even if I'd wanted to.

I got up stiffly and switched on the light. I was beginning to regret agreeing to help Tom with the examination of the body from the cabin. *A shower and breakfast. Things will look better then.*

I spent fifteen minutes running through exercises to strengthen my abdominal muscles, then went into the bathroom and turned on the shower. I turned my face up to the hot spray, letting the needles of water sluice away the lingering effects of the dream.

By the time I emerged, the last vestiges of sleep had been washed away. There was a coffee maker in the room, so I set it going as I dressed and powered up my laptop. It would be late morning in the UK, and I sipped black coffee while I checked my emails. There was nothing urgent; I replied to the ones I needed to and left the rest for later.

The restaurant downstairs had opened for breakfast, but I was the only customer. I passed on the waffles and pancakes and opted for toast and scrambled eggs. I'd been hungry when I went in, but even that seemed too much for me, and I managed less than half. My stomach was knotted, though I didn't know why it should be. I'd only be helping Tom with something I'd done myself countless times before, and in far worse circumstances than this.

But telling myself that didn't make any difference.

By the time I went outside the sun was coming up. Although the car park was still in shadow, the deep blue of the sky was paling, shot through with dazzling gold on the horizon.

The hire car was a Ford, the subtle differences in style and automatic transmission a further reminder that I was in another country. Although it was still early, the roads were already busy. It was a beautiful morning. Built-up as Knoxville was, this part of East Tennessee was still lush and verdant. The spring sun hadn't yet developed the shirt-sticking heat and humidity of high summer, and at this time of day the air held an early morning freshness, unsullied by traffic fumes.

It was an easy twenty-minute drive to UT Medical Center. The morgue was located in a different part of the campus from the facility, but I knew my way there from previous trips.

The man on the morgue reception was so huge he made the desk look like a child's toy. He was quilted with so much flesh that he seemed virtually boneless, the strap of his watch digging into the dimpled wrist like cheese wire into dough. His breath came in a faintly adenoidal wheeze as I explained who I was.

'Autopsy suite five. Through the door and down the corridor.' His voice was incongruously high-pitched for such a big frame. He gave a cherubic smile as he handed me an electronic pass card. 'Cain't miss it.'

I swiped the card on the door and went into the morgue itself. The familiar olfactory punch of formaldehyde, bleach and disinfectant greeted me. Tom was already in the tiled autopsy suite, dressed in surgical scrubs and a rubber apron. A portable CD player stood on a bench nearby, quietly playing a rhythmic drum track I didn't recognize. Another, similarly dressed man was with him, hosing down the body that lay on the aluminium table to sluice off the insects and blowfly larvae.

'Morning,' Tom said brightly as the door swung shut behind me.

I tipped my head towards the CD player. 'Buddy Rich?'

'Not even close. Louie Belson.' Tom straightened from the dripping wet chest cavity. 'You're early.'

'Not as early as you.'

'I wanted to get the body X-rayed and send the dental plates over to the TBI.' He gestured to the younger man who was still hosing down the body. 'David, this is Kyle, one of the morgue assistants. I've had him helping out till you got here, but don't tell Hicks.'

Morgue assistants were employed by the Medical Examiner's office, which meant that Hicks was technically Kyle's boss. I'd forgotten that the pathologist was based here, and I didn't envy anyone working for him. Not that it seemed to bother Kyle. He was tall, with a heavy-boned build that was just on the right side of plump. His pleasant moon face beamed from under an untidy mop of hair.

'Hi,' he said, raising a gloved hand.

'One of my students is going to be lending a hand, as well,' Tom went on. 'It doesn't really need three of us, but I promised I'd let her help out on my next examination.'

'If you don't need me here . . .'

'There's going to be plenty to do. It just means we'll finish sooner.' Tom's smile said I wasn't getting away that easily. 'Scrubs and the rest are in the locker room down the corridor.'

I had the changing room to myself. Putting my own clothes in a locker, I pulled on surgical scrubs and a rubber apron. What we were about to do was perhaps the grimmest part of our work, and certainly one of the messiest. DNA tests could take up to eight weeks, and fingerprints only provided an identity match if the victim's were already on record. But even with badly decomposed bodies such as this, the victim's identity and sometimes also the cause of death could be gleaned from the skeleton itself. Before that could be done, though, every last trace of soft tissue had to be removed.

It wasn't a pleasant job.

When I went back to the autopsy suite I paused outside. I could hear Tom humming along to the jazz over the sound of running water. *What if you make another mistake? What if you can't do this any more?*

But I couldn't afford to think like that. I opened the door and went in. Kyle had finished hosing down the body. Dripping water, the dead man's remains glistened as though they had been varnished.

Tom was at a trolley of surgical instruments. He picked up a pair of tissue scissors and pulled the bright overhead light closer as I went over.

'OK, let's make a start.'

The first dead body I saw was when I was a student. It was a young woman, no more than twenty-five or six, who had been killed in a house fire. She'd asphyxiated from the smoke, but her body was untouched by the flames. She was lying on a cold table under the mortuary's harsh, revealing light. Her eyes were partly open, slits of dull white showing between the lids, and the tip of her tongue was protruding ever so slightly from between bloodless lips. What struck me was how *still* she looked. As frozen and motionless as a photograph. Everything she'd done, everything she'd been and hoped to be, had come to an end. For ever.

The realization hit me with physical force. I knew then that no matter what I did, how much I learned, there would always be one mystery I couldn't explain. But in the years that followed that only increased my determination to solve the more tangible puzzles that lay within my scope.

Then Kara and Alice, my wife and six-year-old daughter, were killed in a car accident. And suddenly such things were no longer academic.

For a time I'd retreated to my original profession of medical doctor, believing that way might bring a measure of peace, if not answers. But I'd only been fooling myself. As Jenny and I had found out to our cost, I couldn't run away from my work. It was what I did, what I was. Or so I'd thought until I'd had a knife thrust into my stomach.

Now I wasn't sure of anything any more.

I tried to put the doubts aside as I worked on the victim's remains. After collecting tissue and fluid samples to send for analysis, I used a scalpel to carefully cut away the muscle, cartilage and internal organs, literally stripping the last vestiges of humanity from the body. Whoever it was, he'd been a big man. We'd need to take more accurate measurements from the skeleton itself, but he was at least six two, and heavily boned.

Not an easy man to overpower.

We worked in near silence, Tom humming absently along to a Dina Washington CD as Kyle wound up the hose and busied himself cleaning the tray where the insects and other detritus from the body were caught after being washed off. I'd begun to lose myself in the work when the double doors to the autopsy suite abruptly swung open.

It was Hicks.

'Morning, Donald,' Tom greeted him pleasantly. 'To what do we owe this pleasure?'

The pathologist didn't bother to reply. The dome of his hairless head gleamed like marble under the bright lights as he glared at Kyle.

'The hell are you doing in here, Webster? I've been looking for you.'

Kyle flushed. 'I was just—'

'He's just finishing up,' Tom put in smoothly. 'I asked him to help out. Dan Gardner wants a report on this as soon as possible. Unless you have any objection?'

Hicks could hardly admit to it if he had. He turned his ire on Kyle again. 'I've got an autopsy this morning. Is the suite ready?'

'Uh, no, but I asked Jason to—'

'I told you to do it, not Jason. I'm sure Dr Lieberman and his *assistant* can manage by themselves while you do what you're paid for.'

It took a second or two to realize he meant me. Tom gave him a thin smile. 'I'm sure we can.'

Hicks gave a sniff, disappointed to be deprived of a confrontation. 'I want everything ready in half an hour, Webster. Make sure it is.'

'Yes, sir. I'm sorry . . .' Kyle said, but the pathologist had already turned away. The heavy door swung shut behind him.

'Well, I'm sure we all feel better for that,' Tom said into the silence. 'Sorry, Kyle. I didn't mean to get you into trouble.'

The younger man smiled, but his cheeks still flamed red. 'That's OK. But Dr Hicks is right. I really ought to—'

The door burst open before he could finish. For a second I

thought Hicks might have come back, but it was a harried-looking young woman who appeared rather than the pathologist.

I guessed she was the student Tom had mentioned would be helping us. She was in her early twenties and wore a faded pink T-shirt over well-worn cargo pants, both stretched by her ample build. The bleached blond hair had been pulled into some sort of order by a red and white polka dot Alice band, and her round glasses gave her an amiably startled appearance. It should have clashed with the steel balls and rings that studded her ears, nose and eyebrows, but somehow didn't. Once you'd got over the initial surprise, the painful-looking array of metalwork seemed to suit her.

Her words were tumbling out in a rush before the door had even swung shut.

'God, I can't believe I'm *late*! I left early so I could stop off at the facility to check my project, but then I totally lost track of time! I'm really sorry, Dr Lieberman.'

'Well, you're here now,' Tom said. 'Summer, I don't think you've met David Hunter. He's British, but don't hold that against him. And this is Kyle. He's been holding the fort till you got here.'

A dazed smile spread across Kyle's face. 'Pleased to meet you.'

'Hi.' Summer beamed, revealing an industrial-looking brace. She glanced across at the body, with interest rather than revulsion. It would have been a shocking sight for most people, but the facility helped prepare students for such grim realities. 'I haven't missed anything, have I?'

'No, he's still dead,' Tom reassured her. 'You know where everything is, if you want to get changed.'

'Sure.' She turned to go out, catching a stainless steel trolley full of instruments with her bag. 'Sorry,' she said, steadying it, before disappearing through the doorway.

A stunned quiet settled over the autopsy suite once more. Tom wore a half-smile. 'Summer's our resident whirlwind.'

'I noticed,' I said.

Kyle was still staring at the door with a shell-shocked expression. Tom gave me an amused glance, then cleared his throat.

'The samples, Kyle?'

'What?' The technician looked startled, as though he'd forgotten we were there.

'You were about to get them packed up for the lab.'

'Oh, right. Sure, no problem.'

With a last hopeful glance at the doors, Kyle gathered up the samples and went out.

'I think it's safe to say our Summer's got an admirer,' Tom said wryly. He turned back to the table and suddenly winced, rubbing his breastbone as though he had trapped air.

'Are you OK?' I asked.

'It's nothing. Hicks is enough to give anyone heartburn,' he said.

But his colour wasn't good. He reached for the tray of instruments and gave a gasp of pain.

'Tom—'

'I'm all right, dammit!' He raised his hand as if to ward me off, then turned it into a gesture of apology. 'I'm fine, really.'

I didn't believe him. 'You've been on your feet since before I got here. Why don't you take a break?'

'Because I don't have time,' he said irritably. 'I promised Dan a preliminary report.'

'And he'll get one. Summer and I can finish off removing the soft tissue.'

He gave a grudging nod. 'Maybe just a few minutes . . .'

I watched him go out, struck by how frail he looked. He'd never been a physically imposing man, but the flesh seemed to have melted from him. *He's getting old.* It was a fact of life. But that didn't make it any easier to accept.

Tom's CD had long since ended, leaving the autopsy suite in silence. From somewhere outside I heard a phone ring. It went unanswered, and finally stopped.

I turned back to the victim's remains. The skeleton was almost completely denuded of flesh by now, leaving only the residual soft tissue to be removed by boiling it in detergent. Since it wasn't practical to immerse the whole skeleton in a huge vat there was another grisly process that needed to be undertaken first.

Disarticulation.

The skull, pelvis, legs and arms would have to be severed, a job requiring both care and brute strength. Any damage to the bone would have to be carefully noted, so it wasn't confused with perimortem trauma. I'd started to remove the skull, painstakingly cutting through the cartilage between the second and third cervical vertebrae, when Summer returned.

In her scrubs and apron she looked less out of place in the morgue, except for the ear and nose piercings. The bleached hair was concealed under a surgical cap.

'Where's Dr Lieberman?' she asked.

'He had to go out.' I didn't enlarge. Tom wouldn't want any of his students to know he was ill.

Summer accepted it. 'You want me to start with the detergent?'

I wasn't sure what Tom had in mind, but that seemed as good an idea as any. We began filling large stainless steel vats with detergent solution and set them heating on gas burners. Although the powerful extractor hood over the burners sucked most of the steam and fumes from the room, the combination of bleach and boiling soft tissue gave off a smell disconcertingly reminiscent of both a laundry and a bad restaurant.

'So you're British?' Summer asked as we worked.

'That's right.'

'How come you're over here?'

'Just a research trip.'

'Don't you have research facilities in the UK?'

'We do, but not like yours.'

'Yeah, the facility's pretty cool.' The big eyes regarded me through the glasses. 'What's it like being a forensic anthropologist over there?'

'Cold and wet, usually.'

She laughed. 'Apart from that. Is it any different?'

I didn't really want to talk about it, but she was only being friendly. 'Well, the basics are the same, but there are a few differences. We don't have as many law enforcement agencies as you do over here.' To an outsider, the number of autonomous sheriff and police departments, let alone state and federal agencies, that operated in the US was bewildering. 'But the main difference is the climate. Unless it's a freakish summer, we tend not to get bodies drying out like you do here. The decomposition's more likely to be a wet one, with more moulds and slime.'

She pulled a face. 'Gross. Ever thought of moving?'

Despite myself I gave a laugh. 'Work in the sun belt, you mean? No, I can't say that I have.' I'd talked about myself as much as I wanted to, though. 'So how about you? What are your plans?'

Summer launched into an animated description of her life so far, her ambitions for the future and how she was working in a bar in Knoxville to raise enough money to buy a car. I said little, content to let her carry on her monologue. It didn't slow her work and the torrent of words was relaxing, so that when Tom returned I was surprised to see that nearly two hours had passed.

'You've made progress, I see,' he said approvingly, coming to the table.

'It's been pretty straightforward.' I didn't ask how he was in front of Summer, but I could see he was feeling better. He waited until she'd returned to the pans bubbling on the gas burners, then beckoned me to one side.

'Sorry I took so long, I've been speaking to Dan Gardner. There's been an interesting development. There aren't any fingerprints on file for Terry Loomis, the guy whose wallet was at the cabin, so they still need us to confirm if this is him.' He gestured towards the remains on the table. 'But they got a result on the print from the film canister.

Belongs to a Willis Dexter, thirty-six-year-old mechanic from Sevierville.'

Sevierville was a small town not far from Gatlinburg, perhaps twenty miles from where the body had been found in the mountain cabin. 'That's good, isn't it?'

'You'd think so,' he agreed. 'They found several other of Dexter's fingerprints at the cabin, as well. One of them on a week-old credit card receipt found in Loomis's wallet.'

All of which suggested that Terry Loomis was the victim and Willis Dexter his killer. But there was something odd about Tom's manner that told me it wasn't that simple. 'So is he in custody?'

Tom took off his glasses and wiped them on a tissue, a quizzical smile playing round his mouth. 'Well, that's the thing. It appears Willis Dexter was killed in a car crash six months ago.'

'That can't be right,' I said. Either the fingerprints couldn't be his or the wrong name must have been put on the death certificate.

'Doesn't seem so, does it?' Tom put his glasses back on. 'That's why we're exhuming his grave first thing tomorrow.'

You're nine when you see your first dead body. You're dressed in your Sunday clothes and ushered into a room where wooden chairs have been set out facing a shiny casket that stands at the front. It's balanced on trestles covered with worn black velvet. A piece of blood-red braiding has come loose on one corner. You're distracted by how it's curled up into an almost perfect figure eight, so that you're almost up to the casket before you think to look inside.

Your grandfather's lying in it. He looks . . . different. His face seems waxy, somehow, and his cheeks have a sunken look, like they do when he forgets to put in his teeth. His eyes are shut, but there's even something not quite right about them, too.

You stop dead, feeling a familiar pressure in your chest. A hand presses into your back, propelling you forward.

'Go on now, take a look.'

66

You recognize the voice of your aunt. But you didn't need any urging to go nearer. You sniff, earning a swift cuff on the head.

'Handkerchief!' your aunt hisses. For once, though, you weren't clearing your nose of its almost permanent drip. Only trying to discern what other odours might be masked beneath the perfume and scented candles.

'Why're his eyes shut?' you ask.

'Because he's with the Lord,' your aunt says. 'Don't he look peaceful? Just like he's asleep.'

But he doesn't look asleep to you. What's in the casket looks like it's never been alive. You stare at it, trying to see exactly what's different, until you're steered firmly away.

Over the next few years the memory of your grandfather's corpse never fails to bring with it the same sense of puzzlement, the same tightness in your chest. It's one of your seminal memories. But it isn't until you're seventeen that you encounter the event which changes your life.

You're sitting on a bench, reading during your lunch break. The book is a translation of St Thomas Aquinas' Summa Theologiae *you stole from the library. It's heavy going and naive, of course, but there's some interesting stuff in it. 'The existence of something and its essence are separate.' You like that, almost as much as you liked Kierkegaard's assertion that 'death is the light in which great passions, both good and bad, become transparent'. All the theologians or philosophers you've read contradict each other, and none of them have any real answers. But they're closer to the mark than the sophomore posturings of Camus and Sartre, who hide their ignorance behind a mask of fiction. You've outgrown them already, just as you're already on your way to outgrowing Aquinas and the rest. In fact you're beginning to think you won't find the answer in any book. But what else is there?*

There've been whisperings at home lately about where the money's coming from to send you to college. It doesn't bother you. It'll come from somewhere. You've known for years that you're special, that you're destined for greatness.

It's meant to be.

You chew and swallow the packed sandwiches mechanically as you read, without enjoyment or taste. Food is fuel, that's all. The most recent operation cured the nasal drip that blighted your childhood, but at a cost. By now your sense of smell is all but burnt out, reducing everything but the spiciest of foods to the blandness of cotton wool.

Finishing the tasteless sandwich, you put the book away. You've just gotten up from the bench when a screech of brakes is followed by a meaty thud. You look up to see a woman in the air. She seems to hang for a moment before crashing down in a sprawl of limbs, almost at your feet. She lies twisted on her back, face tilted to the sky. For a second her eyes meet yours, wide and startled. There's no pain or fear in them, only surprise. Surprise and something else.

Knowledge.

Then the eyes dull and you know instinctively that whatever force had animated the woman has gone. What lies at your feet now is a sack of meat and broken bone, nothing more.

Dazed, you stand there as other people crowd round the body, jostling you aside until it's screened from view. It doesn't matter. You've already seen what you were meant to.

All that night you lie awake, trying to recall every detail. You feel breathless and shaken, on the verge of something immense. You know you've been given a glimpse of something momentous, something both everyday and profound. Except that for some reason the woman's face, the eyes that seemed to burn into yours, now maddeningly elude you. You want – no, you need – to see that moment again in order to understand what happened. But memory isn't up to the task, any more than it was when you stared into your grandfather's casket. It's too subjective; too unreliable. Something this important demands a more clinical approach.

More permanent.

Next day, withdrawing every cent of your college savings, you buy your first camera.

6

Dawn was just a pale band on the horizon when we set off for the cemetery. The sky was still dark, but the stars were slowly disappearing as they were overtaken by the new day. The landscape on either side of the highway was starting to take form, emerging from the darkness like a photograph in a developing tray. Beyond the stores and fast food restaurants, the dark bulk of the mountains rose up as though to emphasize the flimsiness of the man-made facade.

Tom drove in silence. For once he wasn't playing any of his jazz CDs, though whether that was because of the early hour or a reflection of his mood I wasn't sure. He'd picked me up from the hotel, but after a wan smile he'd said little. No one looks their best at that time of day, but there was a greyness to his face that seemed to have nothing to do with lack of sleep.

You probably don't look so good yourself. I'd lain awake into the early hours the night before, apprehensive about what lay ahead. Yet it was hardly my first exhumation, and certainly not the worst. Years before I'd worked on a mass war grave in Bosnia where entire families had been buried. This wouldn't be anything like that, and I knew Tom was doing me a favour in asking me along. By rights I should have jumped at the chance to take part in a US investigation.

So why wasn't I more enthusiastic?

Where I'd once felt confidence and certainty, now there were only doubts. All my energy, the focus I used to take for granted, seemed to have bled out of me on to the floor of my hallway the year before. And if I felt like this now, what would it be like when I was back in the UK, working on a murder inquiry by myself?

The truth was I didn't know.

The eastern horizon was streaked with gold as Tom turned off the highway. We were heading for the suburbs on the eastern fringe of Knoxville, an area I wasn't familiar with. The neighbourhood was a poor one: streets of paint-peeling houses with overgrown and junk-filled front yards. The reflective eyes of a cat gleamed in our headlights as it broke off from eating something in the gutter to glare at us warily as we drove past.

'Not far now,' Tom said, breaking the silence.

After another mile or so the houses began to give way to scrubland, and not long after that we came to the cemetery. It was screened from the road by pine trees and a tall, pale brick wall. A wrought-iron sign proclaimed *Steeple Hill Cemetery and Funeral Home* in an arch above the gates. Cresting it was a stylized angel, its head piously bowed. Even in the half-light I could see that the metal was rusted, its paint flaking.

We drove through the open gates. Gravestones marched along in rows on either side, most of them overgrown and unkempt. They were set against a backdrop of darkly oppressive pine woods, and up ahead I could make out the outline of what must have been the funeral home itself: a low, industrial-looking building topped with a squat steeple.

Off to one side a cluster of parked vehicles announced our destination. We parked by them and climbed out. I shoved my hands in my pockets, shivering in the early morning chill. Mist hung over the dew-silvered grass as we made our way towards the centre of activity.

Screens had been erected in front of the grave, but at that time of day there was no one to see it anyway. A small excavator chugged and juddered as it lifted out another scoop of raw earth, clods dripping from the shovel as it deposited the soil on a growing pile. The air smelled of loam and diesel fumes, but the grave had been almost dug out, a gaping black wound in the turf.

Gardner and Jacobsen stood among a handful of officials and workmen who waited as the excavator cleared another load of earth. Standing slightly apart from them was Hicks. The pathologist's bald head protruded from an oversized mackintosh that made the resemblance to a turtle more striking than ever. His presence was little more than a formality, since the body would almost certainly be handed over to Tom for examination.

It was obvious from his face that he wasn't happy about it.

Another man stood nearby. He was tall and smartly dressed, wearing a camel hair coat over a sombre black suit and tie. He watched the excavator's progress with an expression that could have been either aloof or bored. When he noticed us he seemed to become more alert, his gaze fixed on Tom as we approached.

'Tom,' Gardner said. The TBI agent's eyes were pouched and bloodshot. By contrast Jacobsen looked as fresh as though she'd had nine hours' undisturbed sleep, her belted mac crisp and immaculate.

Tom smiled but said nothing. Slight as the hill was, I could see that he'd been winded by the short walk up from the car. Hicks gave him a jaundiced look but didn't offer any greeting. Ignoring me altogether, he took a grubby handkerchief from his pocket and loudly blew his nose.

Gardner indicated the tall man in the camel hair coat. 'This is Eliot York. He's the owner of Steeple Hill. He helped organize the exhumation.'

'Always glad to assist.' York hurried forward to shake Tom's hand. 'Dr Lieberman, it's an honour, sir.'

The reek of his cologne cut through even the diesel fumes from

71

the excavator. I'd have put him in his late forties, but it was hard to tell. He was a big, fleshy man, with the sort of unlined features that seem to grow heavier instead of ageing. But his dark hair had a matt look that suggested it was dyed, and when he turned I saw it had been carefully brushed to conceal a bald spot on his crown.

I noticed that Tom detached his hand as soon as possible before introducing me. 'This is my colleague, Dr Hunter. He's visiting us from the UK.'

York offered me a perfunctory greeting. Up close the cuffs of the camel hair coat were worn and frayed, and from what I could see of it underneath, his black suit needed cleaning. Judging by the bloodied nicks and tufts of missed whiskers he'd shaved either hurriedly or with a blunt razor. And even his eye-wateringly strong cologne couldn't disguise the cigarette breath or the yellow nicotine stains on his fingers.

He was already turning back to Tom before he'd even released my hand. 'I've heard a lot about your work, Dr Lieberman. And your facility, of course.'

'Thank you, but it isn't exactly "my" facility.'

'No, of course. A credit to Tennessee, though, all the same.' He gave an unctuous smile. 'Not that I'd compare my, ah, *vocation* to yours, but in my own small way, I like to think I'm also carrying out a public service. Not always pleasant, but a necessary one, all the same.'

Tom's smile never wavered. 'Quite. So you carried out this burial?'

York inclined his head. 'We had that honour, sir, although I'm afraid I can't recollect much in this particular instance. We carry out so many, you understand. Steeple Hill provides a fully comprehensive funeral service, including both cremation and interment in this beautiful setting.' He gestured around the unkempt grounds as though they were a stately park. 'My father founded the company in 1958, and we've been serving the bereaved ever since. Our motto is "Dignity and comfort", and I like to think we uphold that.'

The sales pitch was met by an embarrassed silence. Tom looked relieved when Gardner stepped in.

'Shouldn't take much longer. We're almost there,' he said. York's smile faded with disappointment as Tom was deftly steered away.

As though to prove Gardner's point, the excavator deposited one last scoop of dirt on the pile and backed away with a final cough of exhaust. A tired-looking man I took to be a public health official nodded to one of the workmen. Wearing protective overalls and mask, he stepped forward and spread disinfectant into the open hole. Disease doesn't always end with the host's death. As well as the bacteria that flourish on decomposing flesh, hepatitis, HIV and TB are just some of the pathogens that the dead can pass on to the living.

A workman in mask and overalls lowered a short ladder into the grave and began to finish exposing the casket with a shovel. By the time he'd attached straps so it could be lifted out, the sky had lightened to a pale blue and the pine forest was casting long shadows across the grass. When the workman climbed out, he and the others stood on either side of the grave and began hauling the casket out in a macabre reversal of a funeral.

The mud-smeared shape slowly emerged, shedding clods of earth. The men set it down on the boards that had been laid beside the grave and quickly backed away.

'Damn! That stinks!' one of them muttered.

He was right. Even where we stood, the stench of putrefaction was fouling the morning air. Wrinkling his nose, Gardner went over and bent to examine the casket.

'The lid's split,' he said, indicating a crack beneath the caking of soil. 'Don't think it's been broken into, just looks like pretty thin wood.'

'That's finest American pine! It's a perfectly good casket!' York blustered. No one took any notice.

Tom leaned over the casket, sniffing. 'Did you say this was buried six months ago?' he asked Gardner.

'That's right. Why?'

Tom didn't answer. 'Odd. What do you think, David?'

I tried not to show my discomfort as all eyes moved to me. 'It shouldn't smell like that,' I said reluctantly. 'Not after only six months.'

'In case you hadn't noticed, that casket's not exactly airtight,' Hicks said. 'Hole like that, what do you expect?'

I hoped Tom would respond, but he seemed intent on studying the casket. 'It's still had six feet of topsoil on top of it. That far underground the decomposition's going to be much slower than it would be on the surface.'

'I wasn't speaking to you, but thanks for pointing that out,' Hicks said, dripping sarcasm. 'I'm sure being *British*, you know all about Tennessee conditions.'

Tom straightened from the casket. 'Actually, David's right. Even if the body wasn't embalmed the decomp shouldn't smell this bad, broken lid or not.'

The pathologist glared at him. 'Then why don't we take a look?' He motioned brusquely to the workmen. 'Open it up.'

'Here?' Tom said, surprised. Normally the casket would have been transported to the morgue before it was opened.

Hicks seemed to be relishing the moment. 'The casket's already breached. If the body's as far gone as you say, I'd rather find out now. I've wasted enough time already.'

I knew Tom well enough to see his disapproval from the slight pursing of his lips, but he said nothing. Until the body had been officially handed over to him, Hicks was still in charge.

Jacobsen objected anyway. 'Sir, don't you think that should wait?' she said to Hicks as he motioned for a workman to open the casket.

The pathologist gave her a predatory smile. 'Are you questioning my authority?'

'Oh, for God's sake, Donald, just open the damn thing if you're going to,' Gardner said.

With a last glower at Jacobsen, Hicks gestured to one of the workmen who was standing by with a power tool. A high-pitched whine shattered the quiet as one by one the casket's screws were removed. I looked across at Jacobsen, but her face gave no sign of her feelings. She must have felt me watching her, because the grey eyes briefly met mine. For a second I had a glimpse of her anger, and then she looked away.

When the last screw had been taken out, another workman joined the first to help lift the lid. It had warped, and there was a slight resistance before it came loose.

'Jesus H. *Christ*!' one of the men exclaimed, averting his head.

The stench that rose from the casket was overpowering, a foully sweet concentration of rot. The workmen hurriedly moved away.

I stepped up beside Tom to take a look.

A filthy white sheet covered most of the remains, leaving only the skull visible. Most of its hair had sloughed off, although a few thin wisps still clung to it like dirty cobwebs. The body had started to putrefy, the flesh seeming to have melted from the bones as bacteria caused the soft tissue to liquefy. In the casket's closed environment, the resulting fluid had been unable to evaporate. Known as coffin liquor, it was black and viscous, matting the cotton shroud that covered the corpse.

Hicks took a glance inside. 'Congratulations, Lieberman. This one's all yours.'

Without a backward glance he set off towards the parked cars. Gardner was looking at the casket's grisly contents with distaste, a handkerchief held over his mouth and nose in a futile attempt to block the smell.

'That normal?'

'No,' Tom said, shooting an angry look after Hicks.

Gardner turned to York. 'Any idea how this could have happened?'

The funeral home owner's face had crimsoned. 'Of course not! And I resent the implication that this is my fault! Steeple Hill can't be held responsible for what happens to the casket once it's buried!'

'Somehow I didn't think it would be.' Gardner beckoned to the workmen. 'Cover it up. Let's get it to the morgue.'

But I'd been looking at the casket's grisly contents more closely. 'Tom, look at the skull,' I said.

He'd still been staring after the pathologist. Now, giving me a questioning glance, he did as I asked. I saw his expression change.

'You aren't going to like this, Dan.'

'Like what?' Instead of answering, Tom looked pointedly at York and the workmen. Gardner turned to them. 'Can you excuse us a minute, gentlemen?'

The workmen went over to the excavator and began lighting up cigarettes. York folded his arms.

'This is my cemetery. I'm not going anywhere.'

Gardner's nostrils flared as he sighed. 'Mr York—'

'I've got a right to know what's going on!'

'We're still trying to establish that ourselves. Now, if you wouldn't mind . . .'

But York wasn't finished. He levelled a finger at Gardner. 'I've given you every cooperation. And I won't be blamed for this. I want it on record that Steeple Hill isn't liable!'

'Liable for what?' Gardner's tone was dangerously mild.

'For anything! For that!' York gestured wildly at the casket. 'This is a respectable business. I've done nothing wrong.'

'Then you've nothing to worry about. Thanks for your help, Mr York. Someone'll be along to talk to you soon.'

York drew breath to protest, but the TBI agent stared him down. Angrily clamping his mouth shut, the undertaker stalked off. Gardner watched him go with the sort of speculative look a cat might give a bird, then turned to Tom.

'Well?'

'You said this was a white male?'

'That's right. Willis Dexter, thirty-six-year-old mechanic, died in a car crash. C'mon, Tom, what have you seen?'

Tom gave me a wry smile. 'David spotted it. I'll let him break the news.'

Thanks a lot. I turned back to the casket, feeling Gardner and Jacobsen's eyes on me. 'Take a look at the nose,' I told them. The soft tissue had rotted away, leaving a gaping triangular hole lined with scraps of cartilage. 'See down at the bottom of the nasal opening, where it joins the bone that holds the upper teeth? There should be a sill right there, like a sharp ridge of bone jutting out. But there isn't; it blends smoothly into the bone underneath. The shape of the nose is all wrong, too. The bridge is low and broad, and the nasal opening itself is too wide.'

Gardner swore under his breath. 'You sure?' he asked, addressing Tom rather than me.

'Afraid so.' Tom clicked his tongue in annoyance. 'I'd have seen it myself if I'd taken time to look. Any of those cranial features would be pretty strong markers of ancestry by themselves. Take all of them together and there isn't much doubt.'

'Doubt about *what*?' Jacobsen said, bewildered.

'The sill of bone David mentioned is a white facial characteristic,' Tom told her. 'Whoever this is, he doesn't have one.'

Jacobsen frowned as that sank in. 'You mean he's black? But I thought Willis Dexter was white.'

Gardner gave an irate sigh. 'That's right.' He stared down at the body in the casket as though it had let him down. 'This isn't Willis Dexter.'

7

The sun was high and bright, dazzling off the glass and paintwork of the other cars on the highway. Even though it wasn't yet noon, the air above the tarmac rippled with heat and exhaust fumes. Up ahead the traffic slowed to a crawl, snarled round the flashing lights of emergency vehicles that were blocking one lane. A new Lexus was skewed across it at an angle, immaculate and sleek from the back, its front end a jagged mess. Some way from it was what had once been a motorbike. Now it was a crumpled mess of engine parts, chrome and rubber. The road surface around it was stained with what could have been oil, but probably wasn't.

As we crept past, waved on by a stone-faced police officer, I saw onlookers crowding a bridge that spanned the highway, leaning on the railing to gawk at the entertainment below. Then it was behind us, and the traffic resumed its usual flow as though nothing had happened.

Tom seemed more his old self on the drive back from the cemetery. There was a sparkle in his eyes that I knew meant he was intrigued by this latest twist. First fingerprints from a murder scene that belonged to a dead man; now the wrong body had been found in his grave. A puzzle like that was milk and honey to him.

'Starting to look like reports of Willis Dexter's demise might have been a little premature, wouldn't you say?' he mused, fingers drumming on the steering wheel to the Dizzy Gillespie track playing on the CD. 'Faking your own death's a hell of an alibi if you can pull it off.'

I pulled my thoughts back from where they'd wandered. 'So who do you think is in the casket? Another victim?'

'I'm not going to jump to conclusions till we know the cause of death, but I'd say so. It's just about possible that someone at the funeral home got the bodies mixed up by mistake, but under the circumstances that doesn't seem likely. No, much as I hate to admit it, I think Irving was right about this being a serial killer.' He glanced across at me. 'What?'

'Nothing.'

He smiled. 'You'd make a lousy actor, David.'

Normally I'd have enjoyed brainstorming with Tom, but lately I seemed to be too busy second-guessing myself. 'I'm probably just being suspicious. But doesn't it seem a little convenient that the fingerprint on the film canister led straight to another victim's body?'

He shrugged. 'Criminals make mistakes like everybody else.'

'So you believe that Willis Dexter might be still alive? That he's the killer?'

'What do you think?'

'I think I'd forgotten how much you enjoy playing devil's advocate.'

He gave a laugh. 'Just exploring the possibilities. For the record, I agree, it does all seem a mite convenient. But Dan Gardner's no fool. He can be an awkward cuss, but I'm glad he's handling the case.'

I hadn't warmed to Gardner, but Tom didn't bestow praise lightly. 'What did you make of York?' I asked.

'Other than wanting to wash my hand after he'd shaken it, I'm not sure.' He looked thoughtful. 'He's hardly a glowing advertisement for his profession, but he didn't seem too worried about the exhumation.

At least, not until he saw the condition of the casket. I don't doubt he'll have some awkward questions to answer, but I can't see him being so blasé if he'd known what we were going to find.'

'Even so, it's hard to imagine how the wrong body could have been buried without someone at the funeral home knowing about it.'

Tom nodded. 'Almost impossible. But I'm still reserving judgement on York for the time being.' He paused to indicate before changing lanes, overtaking a slow-moving mobile home. 'Nice work back there, by the way. I hadn't noticed the nasal cavity.'

'You would have if you hadn't been so mad at Hicks.'

'Being mad at Hicks is an occupational hazard. I should be used to it by now.' His smile faded as he saw my face. 'OK, out with it. What's bothering you?'

I hadn't planned on bringing it up, but there was no point ducking the issue any longer. 'I don't think my coming here was such a good idea. I appreciate what you're doing, but . . . Well, let's face it, it isn't working out. I think I should go back.'

Until that moment I hadn't even been aware I'd made the decision. Now it seemed as though all my doubts had crystallized, forcing me to accept what I'd been avoiding so far. Yet part of me felt shocked at the admission, knowing there was something irrevocable about it. If I left now I wouldn't be simply cutting my trip short.

I'd be giving up.

Tom was silent for a while. 'This isn't only about what happened at the cabin, is it?'

'That's part of it, but no.' I shrugged, struggling to put it into words. 'I just feel this was a mistake. I don't know, perhaps it was too soon.'

'Your wound's healed, hasn't it?'

'I didn't mean that.'

'I know.' He sighed. 'Can I be frank?'

I nodded; I didn't trust myself to speak.

'You tried running away once before and it didn't work. What makes you think it'll be any better this time?'

I felt my cheeks burn. *Running away?* Was that how he saw it? 'If you mean when Kara and Alice died, then yes, I suppose I did run away,' I said, my voice harsh. 'But this is different. It's like something's missing, and I don't know what.'

'So it's a crisis of confidence.'

'If you like, yes.'

'Then I'll ask you again: exactly how is running away going to help?'

This time it was my turn to fall silent.

Tom didn't take his gaze from the road. 'I'm not going to insult you by giving you a pep talk, David. If it's what you really feel you should do, then leave by all means. I think you'd regret it, but it's your choice. But will you do something for me first?'

'Of course.'

Tom adjusted his glasses. 'I haven't told anyone this except Mary and Paul. But I'll be retiring at the end of the summer.'

I looked at him in surprise. I'd thought he was staying on till the end of the year. 'Is this because of your health?'

'Let's just say I've promised Mary. The point is you were one of my best students, and this is the last chance we're going to have to work together. I'd consider it a great favour if you gave it another week.'

I sat there for a moment, admiring how neatly he'd trapped me. 'I walked into that, didn't I?'

He smiled. 'Yes, you did. But you can hardly break your word to an old man, can you?'

I had to laugh. Oddly enough, I felt lighter than I had done in ages. 'OK, then. A week.'

Tom gave a satisfied nod. He tapped his fingers in time to the trumpet coming from the car speakers.

'So what do you think of Dan's new helper?'

I looked through the window. 'Jacobsen? She seems keen enough.'

'Mm.' The fingers continued to beat out a gentle tattoo on the steering wheel. 'Attractive, wouldn't you say?'

'Yes, I suppose so.' Tom said nothing. I felt my face start to burn. 'What?'

'Nothing,' he said, grinning.

Tom had called ahead to warn the morgue that the exhumed remains were on their way. They'd have to be examined in a separate autopsy suite in order to avoid cross-contamination with the body from the cabin. Just the possibility of that could cause an evidentiary nightmare when the killer was caught.

Assuming he was.

Kyle was talking to two other assistants in the corridor when we arrived. He broke off to take us to the suite he'd prepared, glancing behind us as though expecting – or hoping – to see someone else. He looked crestfallen when he realized there was no one there.

'Is Summer coming in today?'

The attempt at nonchalance wasn't successful. 'Oh, I dare say she'll be stopping by later,' Tom told him.

'Right. I just wondered.'

Tom kept a straight face until Kyle had left the autopsy suite. 'Must be spring,' he said with a smile. 'Gets the sap rising in everyone.'

The casket from Steeple Hill was brought in just as we'd finished changing into scrubs and rubber aprons. It had been transported in a box-like aluminium container; one coffin nestling inside another like Russian dolls. Before anything else the body had to be X-rayed, so Kyle wheeled the whole thing into the radiography room on a trolley.

'Need a hand with this?' he asked.

'No, thanks, we'll manage.'

'Tom . . .' I said. The remains would have to be lifted from the casket to be X-rayed. Decomposition had reduced the body mass, but I didn't want him exerting himself.

He gave an exasperated sigh, knowing what I was thinking. 'We can wait till Summer gets here. I've already gotten Kyle in trouble once.'

'Oh, it's all right. Martin and Jason can cover for me.' Kyle had perked up at the mention of Summer. He gave a shy grin. 'Besides, Dr Hicks isn't here right now.'

Tom reluctantly conceded. 'Well, OK, then. You can help David lift the body out once we've taken photographs.' Just then his phone rang. He looked at its display. 'It's Dan. I better take it.'

While Tom went into the corridor to speak to Gardner, Kyle and I unsnapped the big clips that held the aluminium lid in place.

'So you're British, huh?' he asked. 'From London?'

'That's right.'

'Wow. So what's Europe like?'

I took a moment to wonder how to answer that as I wrestled with a difficult clip. 'Well, it's pretty varied, really.'

'Yeah? I'd like to go someday. See the Eiffel Tower, places like that. I've travelled around the States, but I've always wanted to go somewhere foreign.'

'You should try it.'

'Not on my pay.' He gave a rueful smile. 'So . . . is Summer going to be a forensic anthro like Dr Lieberman?'

'I imagine that's the plan.'

He kept his attention on unfastening the clips, trying to seem unconcerned. 'Does that mean she'll be staying in Tennessee?'

'Why don't you ask her?'

The look he gave me was terrified. He quickly dropped his gaze. 'Oh, no, I couldn't. I just, you know. Wondered.'

I managed not to smile. 'I expect she'll be here for a while yet, anyway.'

'Right.'

He nodded, furiously, burying his head in his work. His shyness

83

was painful to see. I'd no idea if Summer would welcome his attention, but I hoped he found the courage to find out.

We were about to lift off the aluminium lid from the container when Tom returned. His expression was sour.

'Don't bother. Dan doesn't want us to touch the body for the time being. Apparently Alex Irving wants to look at it *in situ*.'

'What for?' I could understand why the profiler had wanted to view the first victim's body in the cabin, but this one was laid out in a coffin. I couldn't see what he hoped to learn from it that he couldn't get from photographs.

'Who knows?' Frustrated, Tom blew out a breath. 'Hicks and Irving in one morning. Lord, this is shaping up to be one hell of a day. And you didn't hear me say that, Kyle.'

The morgue assistant smiled. 'No, sir. Anything else I can do?'

'Not right now. I'll give you a call when Irving gets here. I'm assured he won't be long.'

But we should have known that Irving wasn't the type to worry over keeping anyone waiting. Half an hour, then an hour, went by, and still he hadn't graced us with his presence. Tom and I occupied ourselves in rinsing and drying the remains from the cabin that had been left in detergent overnight. It was nearly two hours before the profiler sauntered into the autopsy suite without knocking. He was wearing an expensive suede jacket over a plain black shirt, his beard little more than a dark shading on the well-fleshed cheeks and softening jaw line.

A girl was with him, pretty and no older than nineteen or twenty. She hung close behind him, as though for protection.

He bestowed an insincere smile upon us. 'Dr Lieberman, Dr . . .' He made do with a vague nod in my direction. 'I expect Dan Gardner told you I was coming.'

Tom didn't return the smile. 'Yes, he did. He also said you'd be here soon.'

Irving raised his hands in mock surrender, giving what I imagine he thought was a disarming grin. '*Mea culpa*. I was about to pre-record a TV interview when Gardner phoned, and it ran late. You know how these things are.'

Tom's face said he knew very well. He looked pointedly at the girl. 'And this is . . . ?'

Irving put a proprietorial hand on the girl's shoulder. 'This is, ah, Stacie. One of my students. She's writing a dissertation on my work.'

'That must be fascinating,' Tom said. 'But I'm afraid she'll have to wait outside.'

The profiler waved a hand, airily dismissing the notion. 'That's OK. I've warned her what to expect.'

'Even so, I'll have to insist.'

The smile became set as Irving locked gazes with Tom. 'I told her she could come with me.'

'Then you shouldn't have. This is a morgue, not a lecture theatre. I'm sorry,' Tom added more gently to the girl.

Irving stared at him for a moment, then gave the girl a regretful smile. 'Looks like I've been overruled, Stacie. You'll have to wait back at the car.'

She hurried out, head bowed with embarrassment. I felt sorry for her, but Irving should have known better than to bring her without first asking Tom. The profiler's smile vanished as soon as the door had closed behind her.

'She's one of my best students. If I'd thought she might embarrass me I wouldn't have brought her along.'

'I'm sure you wouldn't, but that wasn't your decision to make.' Tom's tone ended the discussion. 'David, would you mind bringing Kyle to the radiology suite, please? I'll show Dr Irving where the changing room is.'

'That won't be necessary. I've no intention of touching anything.' The profiler's manner had ice on it now.

'Maybe not, but we're pernickety about things like that. Besides, I'd hate you to get your jacket stained.'

85

Irving glanced down at his expensive suede jacket. 'Oh. Well, perhaps you're right.'

Tom gave me a quick smile as I went out. By the time I'd found Kyle he and Irving were already in the radiography room, standing in silence on opposite sides of the aluminium box containing the casket.

Irving had put on a lab coat over his clothes. He wore a pained expression, massaging either side of his nose with a gloved thumb and forefinger as Kyle and I began to lift the container's lid.

'I hope this won't take long. I have rhinitis and the air condition-ing makes my sinuses – *God!*'

He hastily stepped back, cupping his hand over his nose as the lid came off and released the stench from inside. But to his credit he recovered quickly, lowering his hand and moving forward again as we opened the actual casket.

'Is, ah, is this normal?'

'The condition of the body, you mean?' Tom shrugged. 'Depends what you mean by normal. The decomp is in keeping with an interred corpse. Just not one that's only been buried six months.'

'I presume you have an explanation?'

'Not yet.'

Irving contrived to look surprised. 'So we've got two bodies, both mysteriously more decomposed than they should be. A pattern of sorts there, I think. And I understand this isn't the grave's rightful owner?'

'That's how it looks. This is a black male. Willis Dexter was white.'

'Someone at the funeral home taking colour blindness to new heights, perhaps,' Irving murmured. He motioned at the filthy cotton sheet that covered everything except the corpse's head. 'Can you . . . ?'

'Just a moment. David, would you mind getting a few shots?'

Using Tom's camera, I took photographs of the body, then Tom nodded for Kyle to remove the sheet. The morgue assistant carefully

86

took hold of the makeshift shroud. The fluids released by decomposition had made it adhere to the body, so that it came free only reluctantly. When he saw what was underneath he stopped, looking uncertainly at Tom.

The corpse was naked.

'Oh, definitely a pattern here,' Irving said, sounding amused.

Tom nodded to Kyle. 'Carry on.'

The assistant pulled aside the rest of the sheet. Irving stroked his beard as he considered the body. It seemed a deliberate affectation to me, but perhaps I was biased.

'Well, leaving aside the, ah, *unclothed* aspect for the moment, a few things are immediately obvious,' he asserted. 'The body's been carefully arranged. Hands folded on the chest in the conventional manner, legs straightened as though this was an ordinary burial. Which it patently wasn't. But the body has been treated with evident respect, which is a clear departure from the first victim. Still, all goes to make life more interesting, doesn't it?'

Not theirs. I could see that Irving's attitude irked Tom as well. 'The body we found in the cabin wasn't the first victim,' he said.

'I'm sorry?'

'Assuming that this individual was murdered, which we can't say for sure until we know the cause of death, then he's been dead a lot longer than the man we found yesterday,' Tom said. 'Whoever this was, he died first.'

'I stand corrected,' Irving said, his smile glassy. 'But that only supports my theory. There's a definite progression. And if this Dexter character faked his own death six months ago, as looks likely, then that's hugely symbolic. I thought at first that the killer might be in denial about his sexuality, sublimating his suppressed sexual urges into violence. But this puts a different slant on things. The first victim was covered in a shroud and buried – hidden away in shame, almost. Now, six months later, the body in the cabin is left on display for the world to see. It's shouting, "Look at me! Look what I've done!"

87

Having "buried" his old self the killer's now coming out of the closet, if you like. And given such a huge shift in the way he treated these two victims, I wouldn't be surprised if there are some interim ones we don't know about.'

He sounded quite excited at the prospect.

'So you still think these are gay killings,' Tom said.

'Almost certainly. This all but confirms it.'

'You seem very confident.' I hadn't meant to get involved, but Irving's manner set my teeth on edge.

'We've got two naked corpses, both male. That does seem to point that way, wouldn't you say?'

'Bodies are sometimes transported nude from the morgue. If there was no family to provide clothes then that's how they'd be buried.'

'So this second naked male body is just coincidence? Interesting theory.' He favoured me with a patronizing smile. 'Perhaps you'd also like to explain why the fingerprint Dexter left on the film canister was smeared with baby oil?'

The surprise I felt was mirrored on Tom's face. Irving feigned dismay.

'Oh, I'm sorry, hadn't Gardner mentioned that? No reason why he should, I suppose. But unless the killer has a penchant for moisturizing, there's only one reason I can think of why he was using baby oil at the cabin.'

He let that hang, making sure the barb was sunk before going on.

'In any event, a sexual motivation would also explain the different racial profiles of the victims – the crucial common denominator isn't their skin colour, it's the fact that they're *men*. No, we're definitely dealing with a sexual predator here, and given the conspicuous absence of this Willis Dexter from his own grave, I'd say he's a pretty likely candidate.'

'From what Dan said, I don't think Dexter had a criminal record or any history of violence,' Tom said.

Irving allowed himself a smug smile. 'The really clever predators never do. They keep themselves concealed, often as respectable members of society, until they either slip up or deliberately reveal themselves. Pathological narcissism isn't an uncommon trait amongst serial killers. They tire of hiding their light under a bushel and decide to flex their muscles in public, as it were. Fortunately, most of them eventually trip themselves up with their own vanity. Like this.'

Irving gestured theatrically at the corpse in the casket. By now he'd adopted an almost lecturing tone, as though Tom and I were a pair of not especially bright undergraduates.

'Given the logistics involved, Dexter couldn't have done this without at the very least the help of someone at the funeral home,' he went on confidently. 'Either Dexter worked there himself – which given his background as a mechanic or whatever is unlikely – or he has an accomplice. A lover, maybe. It's possible they might even be working as a team; one dominant and one submissive. Now that really *would* be interesting.'

'Fascinating,' Tom murmured.

Irving gave him a sharp look, as though only now suspecting that his pearls were being wasted on swine. But we were deprived of whatever other insights he might have shared with us by Summer's entrance.

She came into the radiography room but stopped when she saw us standing around the casket. 'Oh! Sorry, shall I wait outside?'

'No need to on my account,' Irving said, favouring her with a broad smile. 'Although I'll defer to Dr Lieberman, of course. He has rather strong views on sheltering students from the facts of life.'

Tom ignored the jibe. 'Summer's one of my graduate students. She's helping us out.'

'Of course.' Irving's smile broadened as he eyed the studs and rings decorating Summer's face. 'You know, I've always been fascinated by body art. I once considered a tattoo myself, but such things are frowned upon in my line of work. But I love the paganistic aspect of piercings, that whole concept of the modern primitive.

So refreshing to find that sort of individualism in this day and age.'

Summer's face bloomed red, but with pleasure rather than embarrassment. 'Thank you.'

'No need to thank *me*.' Irving's charm was on full wattage. 'I have one or two textbooks on primitive body art you might find interesting. Perhaps—'

'If that's all, Professor Irving, we need to make a start here,' Tom interrupted.

Annoyance flickered behind Irving's smile for a moment. 'Of course. Nice meeting you, Miss . . .'

'Summer.'

Irving showed his teeth again. 'My favourite season.'

Peeling off his gloves, he glanced round for somewhere to put them. Failing to find anywhere suitable, he held them out for Kyle to take. The young morgue assistant looked startled, but meekly accepted them.

With a last smile at Summer, Irving went out. There was a hush after the door closed behind him. Summer's face was dimpled in a smile, cheeks blushed crimson beneath the bleached blond hair. Kyle looked crestfallen, the profiler's gloves still dangling from his hand.

Tom cleared his throat. 'So where were we . . . ?'

While I took more photographs of the uncovered remains, he went out to call Gardner. A forensic team would need to examine the casket, but usually that wouldn't happen till after we'd removed the body. The fact that it was naked probably wouldn't alter anything, but I didn't blame Tom for checking with the TBI agent first.

Kyle lingered in the radiography suite, even though there was no real reason for him to be there any more. But seeing the way he looked at Summer I hadn't the heart to tell him he wasn't needed. His expression put me in mind of a kicked puppy.

Tom wasn't long. He came back, his manner brisk. 'Dan says to go ahead. Let's get the body out.'

I started towards the container, but Tom stopped me. 'Kyle, would you mind helping Summer?'

'Me?' The assistant's face turned crimson. He shot a quick glance towards her. 'Oh, uh, sure. No problem.'

Tom gave me a wink as Kyle went to join Summer by the aluminium container.

'Shouldn't you have a bow and arrow?' I murmured, as they prepared to lift the body.

'Sometimes you have to help these things along.' His smile faded. 'Dan's keen to get things moving. Normally I'd leave these remains till I'd finished working on the ones from the cabin, but as things stand—'

There was a sudden exclamation. We looked over to see Kyle straightening beside the casket, staring at one of his gloved hands.

'What's wrong?' Tom asked, going over.

'Something pricked me. When I touched the body.'

'Has it broken the skin?'

'I'm not sure . . .'

'Here, let me see,' I said.

The gloves were heavy duty rubber gauntlets that reached almost to the elbow. Kyle's was slimed with fluids from the decomposing body, but the jagged hole on its palm was clearly visible.

'It's fine, really,' Kyle said.

I took no notice as I pulled off his thick glove. Kyle's hand was wrinkled and pale from being in the rubber. In the centre of his palm was a dark smear of blood.

'Let's get it under the tap. Is there a first aid kit?' I asked.

'There should be one in the autopsy suite. Summer, can you go and get it?' Tom said.

Kyle allowed me to lead him to the sink. I put his hand under the fast-flowing cold water, washing off the blood. The wound was tiny, little more than a pinprick. But that made it no less dangerous.

'Is it OK?' he asked, as Summer returned with the first aid kit.

'If you've had all your shots I'm sure you'll be fine,' I said, putting as much confidence into it as I could. 'You have *had* all your shots?'

He nodded, watching anxiously as I cleaned the wound with antiseptic. Tom had gone over to the casket.

'Whereabouts did you touch the body?'

'It was, uh, the shoulder. The right one.'

Tom leaned closer to look, but didn't touch the corpse himself. 'There's something there. Summer, can you hand me the forceps?'

He reached down and took hold of whatever was embedded in the putrefying flesh. With a little gentle tugging it came free.

'What is it?' Kyle asked.

Tom's expression was studiedly neutral. 'Looks like a hypodermic needle.'

'A *needle*?' Summer exclaimed. 'Omigod, he stabbed himself on a *needle* from that?'

Tom shot her an angry look. But the same thing was going through all our minds. As a morgue worker Kyle would have been immunized against some of the diseases that could be carried by cadavers, but there were others for which there was no protection. Normally, provided care was taken, there was little risk.

Unless you had an open wound.

'I'm sure there's nothing to worry about, but we better get you to the Emergency Room all the same,' Tom said, outwardly calm. 'Why don't you get changed and I'll see you outside?'

Kyle's face had gone white. 'No, I – I'm OK, really.'

'I'm sure you are, but let's get you checked out just to make sure.' His tone didn't leave room for argument. Looking dazed, Kyle did as he'd been told. Tom waited until the door had closed behind him. 'Summer, are you absolutely certain you didn't touch anything?'

She nodded quickly, still pale herself. 'I didn't have the chance. I was going to help Kyle lift the body when he . . . God, do you think he'll be OK?'

Tom didn't answer. 'You might as well get changed too, Summer. I'll let you know if I need you for anything else.'

She didn't argue. He put the needle into a small glass sample jar as she went out.

'Do you want me to go with Kyle?' I asked.

'No, it's my responsibility. You carry on with the other remains for the time being. I don't want anyone going near the casket again until I've X-rayed the body myself.'

He looked as grim as I'd ever seen him. It was possible that the hypodermic needle had snapped off and become embedded by accident, but it didn't seem likely. I wasn't sure what was more disturbing: the idea that the needle had been deliberately planted, or what that implied.

That someone expected the body to be dug up.

Your first time was a woman. More than twice your age and drunk. You'd seen her in a bar, so alcohol-addled she could barely sit still. She'd slipped and swayed on her bar stool, blowsy and overblown, face haggard and red, cigarette burning down to her tobacco-stained finger ends. When she'd thrown her head back and guffawed at the flickering TV screen above the bar, her phlegmy laugh had sounded like a siren call.

You'd wanted her right away.

You'd watched from across the room, your back to her but your eyes never leaving her reflection in the mirror. Swathed in cigarette smoke, she'd approached most of the men in the bar, draping a wattled arm around them in drunken invitation. Each time you'd tensed, jealousy burning like acid in your guts. But each time the arm had been shrugged off, the advances rebuffed. She'd return unsteadily to her stool, loudly demanding another drink to drown her disappointment. And your nervousness would increase, because you knew this was going to be the night.

It was meant to be.

You'd bided your time, waiting until she'd exhausted the barkeeper's patience. You'd slipped out unnoticed while she'd still been screaming at him, obscenities alternating with maudlin entreaties. Outside you'd turned up your collar and hurried to a nearby doorway. It had been fall and a rain-mist had fogged the streets, cloaking the streetlights with yellow penumbras.

You couldn't have asked for a better night.

It had taken longer for her to appear than you'd expected. You'd waited, shivering from cold and adrenalin, nerves beginning to eat away at your anticipation. But you'd held firm. You'd put this off too often already. If you didn't do it now you were frightened you never would.

Then you'd seen her emerge from the bar, her gait unsteady as she tried to pull on a coat that was too thin for the season. She'd walked right past the doorway without noticing you. You'd hurried after her, your heart rapping a staccato counterpoint to your footsteps as you trailed her down the deserted streets.

When you saw the glow of a bar sign up ahead you knew the time had come. You'd caught up to her, fallen in step at her side. You'd planned to say something, but your tongue was thick and useless. Even then she'd made it easy for you, peering around in bleary surprise before the too-red mouth cracked open with a cigarette chuckle.

Hey, lover. Wanna buy a girl a drink?

You had a van parked a few blocks away, but you couldn't wait. When you drew level with the black maw of an alleyway, you'd shoved her into it, trembling as you pulled out the knife.

After that, it had been all fumble and confusion, the quick penetration followed by a rush of fluid. It was over too soon, finished before it had really begun. You'd stood over her, panting, the excitement already starting to turn to something grey and flat. Was that it? Was that all there was to it?

You'd run from the alleyway, chased by disgust and disappointment. It was only later, when your head had started to clear, that you'd begun to analyse where you'd gone wrong. You'd been too eager, in too much of a hurry. These things needed to be done slowly; to be savoured. *How else could you hope to learn anything? In all the rush you hadn't even had a chance to bring the camera from beneath your coat. And as for the knife, you thought, remembering the suddenness of it all . . .*

No, the knife was definitely wrong.

You've come a long way since then. You've refined your technique, honed your craft into an art form. You know now exactly what it is you want, and what you have to do to get it. Still, you look back on that clumsy attempt in the alleyway with something like affection. It had been your first time, and first times were always a disaster.

Practice makes perfect.

8

'Thir*teen*?'

Gardner picked up a sample jar from the collection on the stain-less steel trolley and held it up to see its contents. Like all the rest it contained a single hypodermic needle taken from the exhumed body, a slender steel sliver encrusted with dark matter.

'We found another twelve,' Tom said. He looked and sounded exhausted, the strain of the day's events clearly visible. 'Most of them were embedded in the soft tissue of the arms, legs and shoulders, where anyone who tried to move the remains would be most likely to take hold.'

Gardner set down the jar again, his world-weary features folded into lines of disgust. He'd come alone, and I'd tried to ignore my dis-appointment when I saw that Jacobsen wasn't with him. The three of us were in an unused autopsy suite, where Tom and I had taken the remains after we'd finished X-raying them. The hypodermic needles had shown up as stark white lines against the greys and blacks. He'd insisted on removing them all himself, declining my offer of help. If he could have lifted the body from the casket by himself as well he would. As it was, he'd checked it thoroughly with a hand-held metal detector before allowing either of us to touch it.

After what had happened to Kyle, he wasn't taking any chances.

The assistant had been sent home after spending all afternoon at Emergency. He'd been pumped full of broad spectrum antibiotics, but neither they nor anything else would be effective against some pathogens the needle might have introduced into his bloodstream. He'd have the results of some tests in a few days, but others would take much longer. It would be months before he'd know for sure if he'd been infected or not.

'The needles had been planted with the points facing outwards, so that whoever handled the body was almost certain to impale themselves,' Tom went on, his face drawn with self-reproach. 'This is my fault. I should never have let anyone else handle the remains.'

'You can't blame yourself,' I said. 'There was no way you could have known what was going to happen.'

Gardner gave me a look that said he still wasn't happy about my presence, but kept his thoughts to himself. Tom had already made it clear that he considered I'd as much right to be there as he had, pointing out that it could just as easily have been me who'd been injured.

If Tom hadn't felt sorry for Kyle it might well have been.

'There's only one person to be blamed, and that's whoever did this,' Gardner said. 'It's lucky no one else was hurt.'

'Try telling that to Kyle.' Tom stared at the specimen jars, his eyes ringed with fatigue. 'Have you got any idea yet whose corpse was in the casket?'

Gardner's eyes flicked to the body lying on the aluminium table. We'd hosed it down thoroughly, washing off the worst of the decompositional fluids before Tom had removed the needles. The smell was nothing like so intense as when the casket had first been opened, but it was there, all the same.

'We're working on it.'

'Someone at the funeral home has to know *something*!' Tom protested. 'What does York have to say about it?'

'We're still questioning him.'

'*Questioning* him? Christ almighty, Dan, never mind that there was the wrong body in the grave, someone stuck *thirteen* hypodermic needles in it while it was at Steeple Hill! How the hell could that have happened without York knowing about it?'

The TBI agent's face had set. 'I don't know, Tom. That's why we're questioning him.'

Tom took a deep breath. 'I apologize. It's been a long day.'

'Forget it.' Gardner seemed to regret his earlier reticence. Some of the tension in the autopsy suite seemed to lift as he leaned against the workbench behind him, rubbing the back of his neck. The bright overhead light bleached what little colour there was from his face. 'York claims to have hired someone called Dwight Chambers about eight months ago. According to him this guy was a godsend; worked hard, eager to learn, didn't mind putting in the hours. Then one day he didn't show up and York says he never saw him again. He insists it was Chambers who oversaw Willis Dexter's funeral, who prepared the body and sealed the casket.'

'And you believe him?'

Gardner gave a thin smile. 'I don't believe anyone, you know that. York's a worried man, but I don't think it's because of the murders. Steeple Hill's a mess. That's why he was so keen to help us, hoping if he was nice we'd go away. By the look of things he's been struggling to keep it afloat for years. Cutting corners, hiring casual workers to keep costs down. No taxes, no medical insurance, no questions asked. The bad news is there aren't any records of who's worked there, either.'

'So is there any proof this Dwight Chambers actually existed?' It wasn't until I'd spoken that I remembered I was only there on sufferance. Gardner looked as though he might refuse to answer, but Tom was having none of it.

'It's a legitimate question, Dan.'

Gardner sighed. 'The funeral home's employees come and go so

98

often that Chambers would only have been one of many. It wasn't easy finding anyone who'd worked there long enough to remember him, but we found two who thought they could. The description they gave was pretty vague but matched the one we got from York. White, dark hair, somewhere between twenty-five and forty.'

'Does that fit Willis Dexter?' I asked.

'It fits half the men in Tennessee.' He absently straightened a box of microscope slides so it was aligned with the edge of the work-bench. Catching himself, he stopped and folded his arms. 'But we're looking into the possibility that Dexter and Chambers might be the same person, and that Dexter was cute enough to preside over his own funeral as well as fake his own death. According to the autopsy report he died from massive head trauma when his car hit a tree. No other vehicle was involved, and there was enough alcohol in his system to kill a horse. It was assumed he just lost control.'

'But?' Tom prompted.

'But . . . the car caught fire. The body was only identified through personal effects. So it's possible that a routine autopsy might have overlooked any racial characteristics. And Dexter didn't have any family, so the funeral was just a formality. Closed casket, no embalming.'

It wouldn't have been the first time a burnt-out car had been used to disguise a corpse's identity. But there were still aspects of this that didn't add up.

Tom obviously thought so too. He looked across at the body lying on the table. 'From what I've seen so far that doesn't look burned to me. How about you, David?'

'I wouldn't say so, no.' Although the decomposition could have disguised it to an extent, the body didn't show any evidence of intense heat. Its limbs weren't drawn up into the boxer's crouch characteristic of fire deaths, and while they could have been forcibly straightened afterwards, I would still have expected to see some outward signs, even so.

'Then maybe it was only superficially burned, just enough to scorch the skin,' Gardner said. 'The fact is that Willis Dexter's still missing, and until we've got proof that he's dead that makes him a suspect.'

I spoke without thinking. 'It doesn't make sense for it to be Dexter.'

'Excuse me?'

Go on. Too late to change your mind now. 'If Dexter wanted everyone to think he was dead, why didn't he arrange it so the body was cremated instead of buried? Why go to all that trouble and then leave a corpse in the casket that obviously wasn't his?'

Gardner's face was stone. 'He might have thought that wouldn't matter if it was burned in the car crash. If not for the fingerprints we found in the cabin it wouldn't have.'

'But whoever put the needles in the body obviously expected — *wanted* — it to be exhumed.'

He studied me, as though debating whether to answer or throw me out. 'I'm aware of that. And in case you're wondering, it's also occurred to us that the fingerprint might have been left deliberately. Maybe Dexter did it himself, or maybe he's buried in another grave at Steeple Hill, and someone's got his hand in an icebox. But until we know one way or the other, then he's going to stay a suspect. That all right by you, Dr Hunter?'

I didn't say anything. I could feel the planes of my face tightening.

'David's only trying to help, Dan,' Tom said, which somehow made it worse.

'I'm sure he is.' Gardner's expression could have meant anything. He stood up to go, then paused, addressing Tom as though I wasn't there. 'One more thing. The X-rays of the body from the cabin match Terry Loomis's dental records. We might not be Scotland Yard, but at least we got an ID on one of the victims.'

He gave Tom a nod as he went to the door.

'I'll be in touch.'

★

The day was nearly over by the time we resumed work. We were badly behind schedule, and it didn't help that there were just the two of us. After what had happened to Kyle, Tom wasn't prepared to let Summer help any more.

'It might be bolting the stable door after the horse has gone, but she's only a student. I don't want anything else on my conscience,' he said. He regarded me solemnly over his glasses. 'I'll understand if you want to back out.'

'What happened to "last chance to work together"?' I joked.

The attempt to lighten his mood failed. He rubbed at his breastbone with the heel of his hand, but stopped when he realized I was watching. 'I didn't know then what I'd be getting you into.'

'You didn't get me into anything. I volunteered.'

Tom took off his glasses and began to clean them. He didn't look at me. 'Only because I asked you to. Maybe it would be better if I asked Paul or one of the others to lend a hand.'

The depth of my disappointment surprised me. 'I'm sure Gardner would be happier.'

That at least raised a smile. 'Dan doesn't have anything against you personally. He just likes to do things by the book. This is a high profile homicide investigation and as ASAC he's under pressure to get results. You're an unknown quantity as far as he's concerned, that's all.'

'I get the feeling he'd like me to stay that way.'

The smile became a chuckle, but it soon faded. 'Look at it from my viewpoint, David. After what happened to you last year . . .'

'Last year was last year,' I said, more forcefully than I'd intended. 'Look, I know I'm only here at your invitation, and if you'd rather bring in Paul or someone else to help out, then fine. But I can't duck and run whenever things get complicated. You said as much yourself. Besides, we've found the needles now. What else can happen?'

Tom stared broodingly down at his glasses, still wiping the lenses even though they must have been spotless by now. I stayed silent, knowing he had to decide for himself. Finally, he put the glasses back on.

'Let's get back to work.'

But the relief I felt was soon crowded out as my doubts returned. I found myself wondering if it wouldn't be better for Paul or one of Tom's other colleagues to step in after all. I hadn't come here to take part in an investigation, and my presence was clearly causing friction with Gardner. Tom was every bit as stubborn as the TBI agent, especially when it came to who he worked with, but I didn't want to make things difficult for him.

Even so, I was reluctant to back out now. Whether it was because of what had happened to Kyle, or just that my professional instincts had finally kicked back into life, something in me had changed. For a long time I'd felt as though an essential part of me had been missing, amputated by Grace Strachan's knife. Now something of the old obsessiveness had begun to stir; the need to get to the truth behind a victim's fate. I might only be assisting Tom, but I still felt I had a stake in the investigation. I was loath to simply walk away.

Unless I wasn't given any choice.

While Tom made a start on reconstructing the skeleton that had been confirmed to be Terry Loomis's in one autopsy suite, I began processing the anonymous body from Willis Dexter's casket in the other. It had been hosed down, but the remaining soft tissue still needed to be stripped from it. I hadn't been at it long when Tom poked his head round the door.

'You might want to take a look at this.'

I followed him down the corridor to the other autopsy suite. He'd arranged the large bones of the arms and legs on the examination table, laying them out in an approximation of their anatomical positions. The other bones would follow one by one, until the entire skeleton had been reassembled; a painstaking but necessary job.

Tom went to where the cleaned skull sat at the top of the table and picked it up.

'Beautiful, aren't they? As perfect an example of pink teeth as I've ever seen.'

Cleaned of any decomposing soft tissue, the pink hue was unmistakable. Something had caused blood to be forced into the pulp of Terry Loomis's teeth, either as he'd been killed or shortly afterwards.

The question was what?

'His head wasn't tilted back far enough for gravity to have caused it,' Tom said, voicing my own thoughts. 'I'd say he'd almost certainly have to have been strangled, except for the amount of blood at the cabin.'

I nodded. Judging from what we'd seen, Terry Loomis had virtually bled out. The only problem was if that had happened, then he shouldn't also have had pink teeth. And while it was possible that the wounds we'd seen on his body had been inflicted post mortem, if that were the case they wouldn't have bled nearly so much. So while there was evidence for both strangulation and stabbing as the cause of death, it couldn't be both. Either one ruled the other out.

So which was it?

'Any sign of cuts to the bone?' I asked. If there were, that might indicate a frenzied attack that would point to the wounds being the cause of death.

'None that I've seen so far.'

'What about the hyoid?'

'Intact. No help there, either.'

If the slender bone that sits around the larynx had been broken, it would have meant that Loomis had almost certainly been strangled. But the opposite doesn't apply. It's a common misconception that strangulation always causes the hyoid to break. For all its delicate appearance it's stronger than it looks, so the fact that Loomis's was undamaged didn't prove anything one way or the other.

Tom gave a tired smile. 'Tricky one, isn't it? Be interesting to see if the body from the casket has pink teeth as well. If it has, then my money's on strangulation, cuts or not.'

'We'll have to wait till the skull's been cleaned to know that,' I said. 'The teeth are pretty rotten, and by the look of it the victim was a heavy smoker. There's too much nicotine staining to tell if there's any other discoloration.'

'Well, I suppose we'll just have to—'

Before he could finish the door to the autopsy suite was flung open and Hicks barged in. His face held an alcohol flush, and even from across the room I could smell the sour odour of wine and onions on his breath. He'd clearly enjoyed a good lunch.

Ignoring me completely, he strode up to Tom, bald head gleaming under the fluorescent lights.

'Who the *hell* do you think you are, Lieberman?'

'If this is about Kyle, I'm sorry—'

'*Sorry?* Sorry doesn't begin to cover it. Use your own damn students, not one of my dieners!' He made the unofficial term for morgue assistant sound like an insult. 'Have you any *idea* of how much this could cost if Webster decides to sue?'

'Right now I'm more worried about Kyle himself.'

'Pity you didn't think of that before. You better pray that needle wasn't infected, because if it was I swear this is going to be on your head!'

Tom looked down. He didn't seem to have either the will or the energy to argue.

'It already is.'

Hicks was about to launch into another attack when he became aware of me watching. He glared at me angrily.

'Got something to say?'

I knew Tom wouldn't thank me for interfering. *Bite your tongue. Don't say anything.*

'You've got gravy on your tie,' I said, before I could stop myself.

His eyes narrowed. Until then I think I'd barely registered with him, other than as an extension of Tom. Now I knew I'd put myself in his sights as well, but I didn't care. The Hickses of this world look for excuses to be outraged. Sometimes it's easier just to let them get on with it.

He nodded thoughtfully, as though promising something to himself. 'This isn't over, Lieberman,' he said, giving Tom a final glare before going out.

Tom waited until the door had shut behind him. 'David . . .' He sighed.

'I know, I'm sorry.'

He gave a chuckle. 'Actually, I think it was tomato soup. But in future—'

He broke off with a gasp, his hand going to his chest. I started towards him but he waved me away.

'I'm all right.'

But it was obvious he wasn't. Fumbling off his gloves, he took a small pill case from his pocket and slipped a small tablet under his tongue. After a moment the tension began to go out of him.

'Nitroglycerin?' I asked.

Tom nodded, his breathing gradually becoming less strained. It was a standard treatment for angina, dilating the blood vessels to allow blood to flow more easily to the heart. His colour was already better, but under the harsh lights of the morgue he looked exhausted as he put the pills away.

'OK, where were we?'

'You were just about to go home,' I told him.

'No need. I'm fine now.'

I just looked at him.

'You're as bad as Mary,' he muttered. 'All right. I'll just clear up . . .'

'I can do it. Go on home. This'll still be here tomorrow.'

It was a sign of how exhausted he was that he didn't argue. I felt a pang of concern as I watched him go out. He looked stooped and frail, but it had been a stressful day. *He'll be better after some food and a good night's sleep.*

I almost made myself believe it.

There wasn't much clearing away to be done in Tom's autopsy suite. After I'd finished I went back to my own, where I'd been working on the remains from the exhumed casket. I wanted to finish denuding them of soft tissue and get them into detergent overnight, but as I was about to start I was overcome by a jaw-cracking yawn. I'd not realized till then how tired I was myself. The wall clock said it was after seven, and I'd been on the go since before dawn.

Another hour. You can manage that. I turned to the remains on the examination table. Tissue samples had been sent off to the lab to provide a more accurate time since death, but I didn't need the results of the VFA and amino acid analysis to know that something here didn't add up.

Two bodies, both more decomposed than they should be. There was a pattern there, I'd agree with Irving about that much. Just not one I could make any sense of. The bright overhead light shone dully on the scratched aluminium of the table as I picked up the scalpel. Partially stripped of its flesh, the body lying in front of me resembled a badly carved joint. I bent to start work, and as I did something registered at the edge of my vision.

Something was snagged in the ear cavity.

It was a brown half-oval, no bigger than a grain of rice. Setting down the scalpel, I picked up a pair of small forceps and gently teased it free from the whorl of cartilaginous tissue. I raised it up to examine it, my surprise growing as I saw what it was. *What on earth . . . ?*

It took me a few seconds to realize that the racing in my chest was excitement.

I started searching round for a specimen jar, and gave a start when there was a rap on the door. I looked round as Paul entered.

'Not disturbing you, am I?'

'Not at all.'

He came over and looked down at the body, eyes professionally assessing its tissue-stripped form. He'd have seen worse, just as I had.

Sometimes it's only when you see someone else's reaction – or lack of it – that you realize how we become accustomed to even the most grotesque sights.

'I just saw Tom. He said you were still working, so I thought I'd see how you were getting on.'

'Still behind schedule. You don't happen to know where the specimen jars are, do you?'

'Sure.' He went to a cupboard. 'Tom wasn't looking so good. Was he OK?'

I wasn't sure how much to say, unsure if Paul knew about Tom's condition. But he must have read my hesitation.

'Don't worry, I know about the angina. Did he have another attack?'

'Not a bad one, but I persuaded him to go home,' I said, relieved I didn't have to pretend.

'I'm glad he pays attention to someone. Usually you can't beat him away with a stick.' Paul handed me a specimen jar. 'What's that?'

I put the small brown object into it and held it up for him to see. 'An empty pupal case. Blowfly, by the look of it. It must have lodged in the ear cavity when we hosed down the body.'

Paul looked at it incuriously at first; then I saw the realization hit him. He stared from the specimen jar to the body.

'This came from the body you exhumed this morning?'

'That's right.'

He whistled, taking the jar from me. 'Now how the hell did that get there?'

I'd been wondering that myself. Blowflies were ubiquitous in our line of work, laying their eggs in any bodily opening. They could find their way into most places, indoors or out.

But I'd never heard of any laying their eggs six feet underground.

I screwed the lid on to the jar. 'The only thing I can think of is that the body must have been left on the surface before it was buried. Did Tom tell you about the decomposition?'

107

'That it was worse than it should have been after six months?' He nodded. 'The casing's empty, so the body must have been left out for at least ten or eleven days for the fly to hatch. And six months ago puts the time of death sometime last fall. Warm and wet, so the body wouldn't mummify like it would in summer.'

It was starting to make sense. Either by accident or design, the body had been left to rot before it was put into the casket, which would explain why it was so badly decomposed. Paul was silent for a moment. I knew what he was thinking, and when he turned to me I saw that his excitement matched my own.

'Is the casket still here?'

We left the autopsy suite and went to the storeroom where the casket and aluminium container were awaiting collection by forensic agents. When we opened it the smell of putrefaction was as bad as ever. The shroud was crumpled inside, clotted and rank.

Using a pair of forceps, Paul drew it open.

Until now it had been the body itself that had commanded everyone's attention, not what it had been wrapped in. Now we knew what to look for, though, they weren't hard to find. More pupal cases lay in the cotton sheet, camouflaged by the viscous black slurry from the corpse. Some were broken and empty, already hatched like the one I'd found, but others were still whole. There were no larvae, but after six months their softer bodies would have long since disintegrated.

'Well, that settles it,' Paul said. 'You might explain away one, but not this many. The body must have been pretty badly decomposed before it was sealed in here.'

He reached for the casket lid, but I stopped him. 'What's that?'

Something else was half hidden in the folds of cotton. Taking the forceps from Paul, I gently teased it free.

'What is it, some kind of cricket?' he asked.

'I don't think so.'

It was an insect of some kind, that much was obvious. Well over an inch in length, it was slender with a long, segmented carapace. It had been partially crushed, and its legs had curled in death, emphasizing the elongated teardrop shape of its body.

I set it down on the sheet. Against the white background, the insect looked even more out of place and alien.

Paul leaned forward for a closer look. 'Never seen anything like that before. How about you?'

I shook my head. I'd no idea what it was either.

Only that it had no right to be there.

I worked for another two hours after Paul left. Finding the unknown insect had blown away any vestiges of my earlier tiredness, so I'd carried on until I'd got all the exhumed remains soaking in vats of detergent. I was still buzzing with adrenalin as I left the morgue. Paul and I had decided not to bother Tom with our discovery that night, but I felt convinced that it was a breakthrough. I didn't know how or why, not yet. But my instincts told me the insect was important.

It was a good feeling.

Still preoccupied, I made my way across the car park. It was late and this part of the hospital was deserted. My car was almost the only one there. Streetlights ran round the edges of the car park, but its interior was in almost total darkness. I was halfway across, starting to reach in my pocket for my car keys, when suddenly the hairs on the back of my neck stood up.

I knew I wasn't alone.

I turned quickly, but there was nothing to see. The car park was a field of darkness, the few other cars there solid blocks of shadow. Nothing moved, yet I couldn't shake the feeling that there was something – someone – nearby.

You're just tired. You're imagining things. I set off for my car again. My footsteps sounded unnaturally loud on the gravelled surface.

And then I heard a stone skitter behind me.

I spun round and was blinded by a bright stab of light. Shielding my eyes, I squinted past it as a dark figure with a torch emerged from behind the tank-like shape of a pick-up truck.

It stopped a few feet away, the torch still directed on to my face. 'Mind tellin' me what you're doing here?'

The voice was gruff and threateningly civil, the accent a heavy twang. I made out epaulettes beyond the torch beam, and relaxed as I realized it was only a security guard.

'I'm going home,' I said. He didn't move the light from my face. Its brightness prevented me from making out anything apart from the uniform.

'Got some ID?'

I fished out the pass I'd been given for the morgue and showed it to him. He didn't take it, just dipped the torch beam on to the plastic card before raising it to my face again.

'Could you shine that somewhere else?' I said, blinking.

He lowered the torch a little. 'Workin' late, huh?'

'That's right.' Blotches of light danced in my vision as my eyes tried to adjust.

There was a throaty chuckle. 'Graveyard shift's a bitch, ain't it?'

The torch beam was switched off. I couldn't see anything, but heard his footsteps crunch away across the gravel. His voice floated back to me from the darkness.

'Y'all drive carefully, now.'

You watch the lights from the car recede, waiting until they've disappeared before you step out from behind the pick-up. Your throat is sore from deepening your voice, and your pulse is racing, either from excitement or frustration, you can't be sure.

The idiot never realized how close he came.

You know you took a chance confronting him like that, but you couldn't help it. When you saw him coming across the car park it seemed a God-given opportunity. There was no one else around, and chances were no one would have missed him till the next day. Without even thinking about it, you dogged his steps from the shadows, closing the distance between you.

But quiet as you were, he must have heard something. He stopped and turned round, and although you could still have taken him if you'd wanted, it made you think again. Even if your foot hadn't stubbed that damn stone, you'd already decided to let him go. Lord knows, you're not afraid to take chances, but some Brit no one's ever heard of wasn't worth the risk. Not now, not when the stakes are so high. Still, you'd been sorely tempted.

If it hadn't been for what you've got planned for tomorrow you might have gone ahead anyway.

You smile as you think of it, anticipation bubbling up inside you. It's going to be dangerous, but no one wins any prizes by playing safe. Shock and awe, that's what you want. You've hidden your light under a bushel for long enough, watched your lessers take all the glory. High time you got the recognition you deserve. And after tomorrow no one's going to be in any doubt what you're capable of. They think they know what they're dealing with, but they've no idea.

You're just getting started.

You take a deep breath of the warm spring night, savouring the sweetness of blossom and the faintly treacly smell of asphalt. Feeling strong and confident, you climb into the pick-up. Time to go home.

You've got a busy day ahead.

9

The last remnants of an early morning mist still hung between the trees bordering the woodland path. Shafts of low sunlight broke through the canopy of new leaves and branches, dappling the ground with a cathedral light.

A lone figure sat reading a newspaper at a picnic bench made from rough-cut pine. The only sound came from the rustle of the pages, and the hollow rattle of a woodpecker in the trees nearby.

The newspaper reader glanced up, idly, as a piercing whistle came from the trail off to the left, where it curved out of sight. A moment later a man appeared. He wore an irritated expression, and was looking into the undergrowth at either side as he walked. He had a dog lead in one hand, the empty chain swinging in rhythm with his brisk steps.

'Jackson! Here, boy! Jackson!'

His calls were interspersed with more whistles. After a single incurious glance, the reader went back to the news headlines. The dog walker paused as he drew level, then came across.

'Haven't seen a dog, have you? A black Labrador?'

The reader glanced up, surprised to have been addressed. 'No, I don't think so.'

The dog walker gave a snort of annoyance. 'Damn dog. Probably off chasing squirrels.'

The reader gave a polite smile before going back to the newspaper. The man with the dog chain chewed his lip as he stared up the trail.

'I'd appreciate it if you'd keep an eye open for him,' he said. 'You see him, don't let him get away. He's friendly, he won't bite.'

'Sure.' It was said without enthusiasm. But as the dog walker looked forlornly around the reader reluctantly lowered the newspaper again.

'There *was* something making a noise in the bushes a while ago. I didn't see what was making it, but it could have been a dog.'

The dog walker was craning his head to see. 'Where?'

'Over there.' The reader gestured vaguely towards the undergrowth. The dog owner peered in that direction, chain swinging loosely in his hand.

'By the trail? I can't see anything.'

With a sigh of resignation, the reader closed the newspaper. 'I suppose it's easier to show you.'

'I appreciate this,' the dog walker said with a smile, as they entered the trees. 'I haven't had him long. Thought I'd gotten him trained, but every now and again he'll just take off.'

He paused to whistle and call the dog's name again. The reader gave the heavy dog chain an uneasy glance, then looked back towards the trail. No one was in sight.

Suddenly the dog walker gave a cry and ran forward. He dropped to his knees by a clump of bushes. The body of a black Labrador lay behind them. Blood matted the dark fur on its crushed skull. The dog walker's hands hovered over it, as though scared to touch it.

'*Jackson?* Oh, my God, look at his *head*, what *happened*?'

'I broke his skull,' the newspaper reader said, stepping up behind him.

The dog walker started to rise, but something clamped round his neck. The pressure was unrelenting, choking off his cry before he could make it. He tried to struggle to his feet, but he was off balance and his arms and legs had no strength. Belatedly, he remembered the dog chain. His brain tried to send the necessary commands to his muscles, but the world had already started to turn black. His hand spasmed once or twice, then the chain dropped from his limp fingers.

High above in the branches, the woodpecker cocked its head to assess the scene below. Satisfied there was no threat, it resumed its hunt for grubs.

Its *rat-a-tat* echoed through the woodland morning.

I woke feeling better than I had in months. For once my sleep had been undisturbed, and the bed looked as though I'd barely moved all night. I stretched, then ran through my morning exercises. Normally it was a real effort, but for once it didn't seem so bad.

After I'd showered I turned on the TV, searching for an international news channel as I dressed. I skipped through one station after another, letting the stream of advertisements and banal chatter wash over me. I'd gone past the local news station before I registered what I'd seen.

Irving's smoothly bearded face reappeared on the screen as I flicked back. He was looking thoughtfully sincere as he spoke to a female interviewer who had the painted-on prettiness of a shop-window dummy.

'. . . of course. "Serial killer" is a phrase that's badly over-used. A true serial killer, as opposed to someone who merely kills multiple victims, is a predator, pure and simple. They're the tigers of modern society, hiding unseen in the tall grass. When you've dealt with as many as I have, you learn to appreciate the difference.'

'Oh, for God's sake,' I groaned. I remembered that Irving had been late at the morgue the day before because he was pre-recording a TV interview, but I hadn't given it much thought. My mood curdled as I watched.

'But it is correct that you've been called in by the TBI to provide an offender profile following the discovery of a mutilated body in a Smoky Mountain rental cabin?' the interviewer persisted. 'And that a second body has been exhumed from a cemetery in Knoxville as part of the same case?'

Irving gave a rueful smile. 'I'm afraid I'm not at liberty to comment on any ongoing investigations.'

The interviewer nodded understandingly, her lacquered blond hair remaining immobile. 'But since you are an expert on profiling serial killers, presumably the TBI are worried that's what they may be dealing with, and that this may be just the start of a killing spree?'

'Again, I'm afraid I really can't comment. Although I'm sure people will draw their own conclusions,' Irving added innocently.

The interviewer's smile revealed perfect white teeth beneath the blood-red lipstick. She crossed her legs. 'So can you at least tell me if you've formed a profile of the killer?'

'Now, Stephanie, you know I can't do that,' Irving said, with an urbane chuckle. 'But what I can say is that all the serial killers I've encountered – and believe me, there have been quite a few – have one defining characteristic. Their ordinariness.'

The interviewer cocked her head as though she'd misheard. 'I'm sorry, you're saying that serial killers are ordinary?' Her surprise was transparently artificial, as though she'd known what he was going to say in advance.

'That's right. Obviously, that isn't how they regard themselves: quite the opposite. But in truth they're nonentities, almost by definition. Forget the glamorous psychopath of popular fiction; in the real world these individuals are sad misfits for whom killing has become the primal urge. Cunning, yes. Dangerous, certainly. But their one defining feature is that they blend into the crowd. That's what makes them so difficult to detect.'

'But surely that also makes them harder to catch?'

Irving's smile widened into a wolfish grin. 'That's what makes my job so challenging.'

The interview ended, cutting to a studio anchorwoman. *'That was behaviouralist Professor Alex Irving, author of the bestselling* Fractured Egos, *speaking yesterday to—'*

I snapped off the set. 'Nothing wrong with his ego,' I muttered, tossing aside the TV remote. There had been no justification for the interview. It had served no purpose except to give Irving the opportunity to preen on TV. I wondered if Gardner had known about it. Somehow I couldn't see him taking kindly to Irving using the investigation to promote his new book.

Still, not even the psychologist's smugness could spoil the anticipation I felt as I drove to the morgue. For once I was there before Tom, but only just. I'd barely changed into scrubs when he arrived.

He looked better than he had the night before, I was relieved to see. Food and a good night's sleep might not cure everything, but they rarely hurt.

'Someone's eager,' he said when he saw me.

'Paul and I found something last night.'

I showed him the pupal cases and the mystery insect, explaining how we'd stumbled across them.

'Curiouser and curiouser,' he said, studying the insect. 'I think you're right about the body decomposing on the surface before it was buried. As for this . . .' He lightly tapped the jar containing the dead insect. 'I haven't a clue what it is.'

'Oh.' I'd assumed Tom would have been able to identify it.

'Sorry to disappoint you. Blowflies and beetles are one thing, but I haven't come across anything like this before. Still, I know someone who should be able to help us. You haven't met Josh Talbot, have you?'

'I don't think so.' I'd met several of Tom's colleagues, but the name didn't ring a bell.

'He's our resident forensic entomologist. The man's a walking insect encyclopaedia. If anyone can tell us what this is, Josh can.'

While he called Talbot I set about rinsing the bones from the exhumed body that had been soaking in detergent overnight. I'd got as far as setting the first of them to dry in the fume cupboard when Tom closed his phone.

'We're in luck. He's about to leave for a conference in Atlanta but he's going to drop by first. Shouldn't take him long.' He began helping me put the bones in the fume cupboard. 'Did you catch our friend Irving on TV last night, by the way?'

'If you mean the interview, no, but I saw it this morning.'

'Lucky you. Must be re-running it.' Tom smiled and shook his head. 'You have to hand it to him, he doesn't miss a chance, does he?'

He'd barely finished speaking when there was a light knock on the door. He frowned. 'Can't be Josh already,' he said, going to open it.

It wasn't. It was Kyle.

Swallowing his surprise, Tom moved aside to let him in. 'I didn't expect to see you back yet. Why aren't you taking some time off?'

Kyle gave a strained smile. 'They offered, but it isn't right that the other guys should have to cover for me. I feel fine. And I guess I'd rather work than sit at home.'

'How's the hand?' I asked.

He held it up to show us. A small plaster on the palm was the only sign of what had happened. Kyle looked at it as though it wasn't part of him. 'Not much to look at, is it?'

There was an awkward silence. Tom cleared his throat. 'So . . . How are you bearing up?'

'Oh, pretty good, thanks. Be a while before I get the test results, but I'm looking on the bright side. The hospital said there're post-exposure treatments for HIV and some other things if I want them. But the way I see it, the body might not even have been infected. And even if it was I might not catch anything, right?'

'You should still consider them, at least,' Tom said. He gestured helplessly. 'Look, I'm sorry about—'

'Don't!' The sharpness showed how much pressure Kyle was under. He gave an embarrassed shrug. 'Please, don't apologize. I was just doing my job. Stuff happens, y'know?'

There was an uncomfortable pause. Kyle broke it.

'So . . . where's Summer?' He did his best to sound casual, but the attempt was no more convincing than before. It wasn't hard to guess the real reason he'd come to see us.

'I'm afraid Summer won't be helping us any more.'

'Oh.' His disappointment was obvious. 'Can I help?'

'Thanks, but David and I can manage.'

'Right.' Kyle nodded emphatically. 'Well, anything you need, be sure to let me know.'

'I will. You take care now.' Tom's smile lasted only until the door had closed. 'Lord . . .'

'He's right,' I said. 'He was doing his job. It's no good blaming yourself. And if it comes down to it, it should have been me helping Summer, not him.'

'It wasn't your fault, David.'

'Or yours either. Besides, we don't know yet that the needle was contaminated. He might be fine.'

It was a faint hope, but no good would come from Tom's torturing himself. He drew himself up.

'You're right. What's done's done. Let's just concentrate on catching this son of a bitch.'

Tom rarely swore, and it was a sign of his agitation that he didn't seem to realize he had. He went to the door, then paused.

'Almost forgot. Mary wanted me to ask if you eat fish.'

'Fish?' The change of tack threw me. 'Yes, why?'

'You're coming over for dinner tonight.' The eyebrows climbed as he enjoyed my discomfort. 'Sam and Paul are coming as well. Don't tell me you'd forgotten?'

It had completely slipped my mind. 'No, of course not.'

He grinned, some of his usual humour returning. 'Perish the thought. Not as though you've had anything else to think about, is it?'

★

118

There are two hundred and six bones in the adult human body. They vary in size from the femur, the heavy thigh bone, to the tiny ossicles of the inner ear, the smallest no larger than a grain of rice. Structurally, the skeleton is a marvel of biological engineering, as intricate and sophisticated as anything designed by man.

Reassembling it isn't a straightforward task.

Stripped of any last vestige of decaying tissue, the bare bones of the man buried in Willis Dexter's casket told their own story. Their African ancestry was now unmistakable, immediately evident in the slightly straighter, lighter bone structure and more rectangular eye orbits. Whoever this was, he'd been of medium height and build, and judging from the wear to his joints he was between his mid-fifties and early sixties. There were long-healed breaks in the right femur and left humerus, both probably the result of childhood accidents, and signs of arthritis on his knee and ankle joints. The damage was more evident on the left than on the right, which meant he had favoured that side when he walked. And the left hip was also badly eroded, the ball and socket pitted and worn. If he hadn't been contemplating hip replacement surgery when he died, then he would have been all but crippled before much longer.

Not that it made any difference to him now.

Like Terry Loomis's, the man's hyoid was still intact. That didn't mean anything either way, but when I lifted the dripping skull from the vat I smiled grimly to myself. The teeth were still brown and stained, but below where the gum had once been a band of clean enamel was now exposed.

There was no mistaking the pink discoloration.

I was still examining the skull when Tom came in. A short, paunchy man in his fifties was with him. His thinning ginger hair was swept half-heartedly over a reddened crown, and he carried a battered leather briefcase that fairly bulged with books.

'Josh, I'd like you to meet David Hunter,' Tom said as he entered.

'David, this is Josh Talbot. What he doesn't know about bugs isn't worth knowing.'

'He knows I hate that word,' Talbot said affably. He was already looking round the room, bright-eyed with anticipation. His gaze lingered on the bones, but not for long. They weren't why he was here.

'So where's this mystery insect you've got for me?'

When he saw the specimen jar his entire face lit up. He bent down to study it at eye level. 'Well, now, this *is* a surprise!'

'You recognize it?' Tom asked.

'Oh, yes. Quite a find, too. There's only one other part of Tennessee where this species of *Odonata* has been confirmed. There've been sightings round here before, but it isn't every day you come across one of these beauties.'

'I'm glad to hear it,' Tom said. 'Do you think you could tell us what it is?'

Talbot grinned. '*Odonata* are dragonflies and damselflies. What you've got here is a dragonfly nymph. A swamp darner, one of the biggest species in North America. They're widespread across most eastern states, although less so in Tennessee. Here, I'll show you.'

He rummaged in his briefcase and produced a thick, dog-eared old textbook. Humming to himself, he set it on the workbench and began flicking through its pages.

He stopped and tapped on one. 'Here we go. *Epiaeschna heros*, the swamp or hero darner, as they're sometimes called. Migratory, generally found by wooded roadsides and ponds in summer and fall, but adults can hatch in spring in warmer regions.'

The page showed a photograph of a large insect shaped like a miniature helicopter. It had the familiar double wings and streamlined body of the dragonflies I'd seen at home, but there the resemblance ended. This one was as long as my finger and almost as thick, its brown body tiger-striped with bright green. But the most striking features were its eyes: huge and spherical, they were a vivid, electric blue.

'I know dragon hunters in Tennessee who'd give their hind teeth to see an adult hero,' Talbot enthused. 'Just look at those eyes! Incredible, aren't they? On a sunny day you can spot them a mile away.'

Tom had been examining the book. 'So what we found is the nymph of one of these?'

'Or naiad, if you prefer.' Talbot steepled his fingers, warming to his theme. 'Dragonflies don't have a larval stage. They lay their eggs in still or slow-moving water, and when the nymphs hatch they're completely aquatic. At least, they are until they mature. Then they crawl out on to a plant or grass stem to metamorphose into an adult.'

'But dragonflies aren't normally attracted to carrion, are they?' I asked.

'Oh, Lord, no.' He sounded shocked. 'They're predators. They're sometimes called mosquito hawks, because that's their main diet. That's why you generally see them near water, although swamp darners are partial to winged termites, too. You say this specimen was found in a casket?'

'That's right. We think it was probably bundled there along with the body,' Tom told him.

'Then I'd say the body had to have been left close to a pond or lake. Probably right by the water's edge.' Talbot picked up the jar. 'When this little fella crawled out to metamorphose it obviously got scooped up as well. Even if it wasn't crushed, burying it in the cold and dark would have killed it.'

'Are there any particular areas where this species is likely to be found?' Tom asked.

'Not in fast-running streams or rivers, but pretty much any woodland where there's standing water. They're not called swamp darners for nothing.' Talbot glanced at his watch, then packed the book back into his briefcase. 'Sorry, have to go. If you find any live specimens, be sure to let me know.'

Tom went to see Talbot out. He returned a few minutes later, his face thoughtful.

'At least we know now what it was we found,' I said. 'And if the body was left near a pond or still water it gives Gardner a little more to go on.'

Tom didn't seem to have heard. He picked up the skull and examined it, but absently, as though he wasn't really aware of what he was doing. Even when I told him about the intact hyoid and pink teeth of the exhumed remains, he still seemed distracted.

'Is everything OK?' I asked at last.

He put down the skull. 'Dan Gardner called just before Josh arrived. Alex Irving's missing.'

My first thought was that there must be some mistake; I'd only seen the profiler on TV that morning. Then I remembered that the interview had been shot the day before: what I'd watched had been a repeat. 'What happened?'

'No one's sure. Apparently he went out early this morning and didn't come back. He hasn't been seen since.'

'Isn't it a bit soon to say he's missing if he's only been gone a few hours?'

'Ordinarily. But he'd taken his dog for a walk.' Tom's eyes were troubled. 'They found it with its skull smashed in.'

The blood swirls down the sink, marbling the fast-flowing cold water with carmine strands. A piece of meat, drained to a pale pink now the blood has been washed from it, catches in the plughole. You jab it with your finger until it's been forced through.

Whistling absently to yourself, you chop fresh chillies and drop them into a pan with a handful of garlic salt. When they've started to sizzle you scoop up the meat and drop that on it as well. The wet flesh spits and hisses when it hits the hot fat, sending up a blast of steam. You give it a quick stir, then leave it to brown. Opening the cold cupboard, you take out a carton of orange juice, cheese and mayonnaise. You select a glass that looks reasonably clean and wipe it with your finger. Dust covers every surface, but you don't notice. If you did you wouldn't care. Occasionally, like a veil lifting, you'll register the dilapidation of your surroundings, the way every corner is furred with the detritus of years, but it fails to bother you. Decay is part of the natural order of things, and who are you to deny nature?

You drink a glassful of orange straight off, wiping your mouth with the back of your hand before you spread mayonnaise on two slices of processed white bread and top it with thick chunks of cheese. Pouring yourself more orange, you go to the big table in the centre of the kitchen. There isn't much room left on it, so you balance your plate on a corner and pull up a chair. The sandwich tastes of nothing, as usual, but it'll fill your stomach. You don't really miss not being able to taste or smell anything, not any more.

Not when there's so much else to savour.

Things are going to move fast now, but that's OK. It's only what you expected, and you're at your best under pressure. Everything's going exactly like you knew it would. Just like you planned it. Leaving everything at the mountain cabin was a risk, but a calculated one. It had felt strange, working out there away from your own environment. The film canister was an inspired move, but leaving the body there for them to find had gone against the grain. Still, it had been necessary. You wanted to make an impact, and how better than to give them a kill site to play with? Let them run themselves ragged trying to guess what you're going to do next. It won't do them any good.

By the time they realize it'll be too late.

You finish the sandwich, washing it down with orange juice that tastes of nothing but cold. A patch of mayonnaise flecks one corner of your mouth as you go to the stove to check the pan. You lift the lid and inhale the sudden belch of steam. You can't smell it but it makes your eyes water, and that's a good sign. The meat is starting to brown nicely. Pork rather than beef, same as always. Cheaper, and it's not like you can tell the difference anyway.

You pick up a spoon and try some. Even though you can't taste anything, it's so heavily spiced that it burns your mouth. Just like a good chilli should. You throw in a couple of cans of tomato, then take the pan off the heat and cover it. It'll cook slowly on its own now, and by the time you get back it'll be just right.

You're a great believer in leaving things to stew in their own juices.

123

You pick up the plastic bag of dirty clothes you need to drop off at the laundry, reminding yourself that you need to stock up on supplies again, too. More cans of tomato, and you're getting low on batteries and flypaper. You examine the sticky strips hanging from the ceiling. At least, they used to be sticky; now they're matted black with dead flies, as well as the husks of larger, more colourful insects.

For a moment a blankness comes over your face, as though the reason for the strips has momentarily escaped you. Then you blink and come back to life. On your way out you pause by the table. The man lying trussed on it looks up at you with terrified eyes, snuffling round the gag in his mouth. You give him a smile.

'Don't you worry, now. I'll be back soon.'

Hoisting the heavy bag of laundry, you go out.

10

Gradually, a picture emerged of what had happened. Irving lived out near Cades Cove, a beauty spot in the foothills of the Smoky Mountains. Each morning before breakfast he would take his dog, a black Labrador, out walking on the trail in the woods behind his home. It was an established part of Irving's routine, and one that he'd mentioned more than once in the profile interviews he was so fond of giving.

At around nine o'clock his PA had let herself into his house, as she did most mornings, and started the coffee percolator, so that Irving's favourite French roast would be ready for him by the time he returned.

Except that this morning he hadn't. The PA – his third in two years – had tried calling his mobile but received no answer. When there was still no sign of him as lunchtime approached, she'd gone out along the trail herself. Less than a half-mile from his house she'd seen a policeman talking to an elderly couple, whose Jack Russell was yapping excitedly on its lead. As she'd passed she'd overheard them telling him about the dead dog that their terrier had found. A black Labrador.

That was when she realized her employer might not be back for lunch after all.

A search of the area revealed a bloodstained steel bar lying near the Labrador's body, and the muddy ground by the dog's body bore evidence of a struggle. But while there were several sets of footprints, none of them were distinct enough for casts.

Of Irving himself, there was no sign.

'We don't know for certain what's happened to him,' Gardner admitted. 'We think all the blood on the bar is from the dog, but until it's been to the lab we can't be sure.'

We were in one of the morgue's offices, down the corridor from the autopsy suites. Windowless and small, it could have belonged to any anonymous business. Gardner had come at Tom's request. This time Jacobsen was with him, cool and unapproachable as ever in a knee-length charcoal grey skirt and jacket. Except for the colour, it looked identical to the blue one I'd seen her in before. I wondered if she had a wardrobe full of identical suits, running the dark spectrum of neutral shades.

Although no one had broached the actual reason for the meeting, we were all aware what it was. Even unspoken it created a palpable tension in the small office. Gardner had restricted his unhappiness at my presence to a disapproving glance. He looked even more careworn than usual, the creases in his brown suit matching those in his face, as though he were subject to a heavier gravity than the rest of us.

'You must have some theories,' Tom said. He sat behind the desk, listening with a brooding expression I knew meant he was biding his time. He was the only one seated. Although there was another chair in front of the desk no one had taken it. The rest of us stayed on our feet, the chair remaining vacant as though awaiting the arrival of a late visitor.

'It's possible Irving was the victim of a random attack, but it's still too soon to say. We're not ruling out anything at this stage,' Gardner said.

Tom's exasperation was beginning to show. 'In that case where's his body?'

'We're still searching the area. For all we know he could have been injured and wandered off. The dog was found in woodland half a mile from the nearest road. That's a long way to carry a grown man, but there's no other way anyone could've got Irving out of there. All we've found so far are footprints and cycle tracks.'

'Then maybe he was forced to walk out himself at gun or knifepoint.'

Gardner's chin jutted stubbornly. 'In broad daylight? Unlikely. But like I said, we're considering every possibility.'

Tom considered him. 'How long have we known each other, Dan?'

The TBI agent looked uncomfortable. 'I don't know. Ten years?'

'It's twelve. And this is the first time you've ever tried to bullshit me.'

'That isn't fair!' Gardner shot back, his face darkening. 'We came here today out of courtesy—'

'Come on, Dan, you know what happened as well as I do! You can't seriously believe it's coincidence that Irving's gone missing the morning after he bad-mouthed a serial killer on TV?'

'Until there's proof I'm not going to jump to conclusions.'

'And what if someone else on the investigation goes missing? Will that be jumping to conclusions too?' In all the years I'd known Tom I'd never seen him so angry. 'Dammit, Dan, one person was injured here yesterday, perhaps seriously, and now this! I have a responsibility to the people working with me. If any of them are at risk then *I want to know about it*!'

Gardner said nothing. He looked pointedly across at me.

'I'll be in the autopsy suite,' I said, heading for the door.

'No, David, you've got as much right to hear this as I have,' Tom said.

'Tom . . .' Gardner began.

'I asked him to help, Dan. If he's going to share the risk he has every right to know what he's got himself into.' Tom folded his arms.

'I'll only tell him what you say anyway, so he might as well hear it from you.'

The two of them stared at each other. Gardner didn't strike me as the type to be easily browbeaten, but I knew Tom wasn't going to budge. I glanced at Jacobsen and saw she looked as uncomfortable as I felt. Then she realized I was watching her, and quickly blanked any hint of emotion from her features.

Gardner gave a resigned sigh. 'Jesus, Tom. All right, it's *possible* there's a connection. But it isn't that simple. Some of Alex Irving's students had complained about his behaviour. Female students. The university'd been turning a blind eye because he was a celebrity professor who could walk into a job anywhere in the state. Then a student accused him of sexual harassment and that opened the flood-gates. The police were brought in, and it looked as though the university was going to cut him loose rather than risk being hit with lawsuits themselves.'

I thought about the blatant way Irving had flirted with Summer and even Jacobsen, despite publicly slapping her down. It didn't surprise me that they weren't the only ones. Evidently not everyone fell for his charm.

'So you think he pulled a vanishing act?' Tom asked doubtfully.

'Like I said, we're considering every possibility. But Irving didn't just have the harassment case hanging over him. The IRS have been investigating him for unpaid tax on all those book deals and TV appearances. He was looking at a bill of over a million dollars, maybe even a jail sentence. He was facing professional and financial ruin no matter what. This might have seemed like an ideal opportunity to get out from under.'

Tom pulled at his lower lip, frowning. 'Even so, killing his own dog?'

'People have done worse for less. And you might as well know, we found a clear set of fingerprints on the bar used to kill Irving's dog. When we ran them we got a match with a petty thief called Noah

Harper. He's a career criminal, with a string of car theft and burglary convictions.'

'If you've got a suspect then why aren't you looking happier?' Tom asked.

'Because for one thing all of Harper's offences in the past have been minor league. And for another he's been missing for nearly seven months. He didn't turn up for his last parole appointment and no one's seen him since. All his belongings were left in his apartment, and the rent was paid up till the end of the month.'

'Is he African American?' I asked. 'Fifty to sixty, with a bad limp?'

It was hard not to enjoy Gardner's surprise. 'How do you know that?'

'Because I think he's in the autopsy suite down the corridor.'

I watched realization put even more folds into his already crumpled face. 'I'm getting slow,' he said, disgusted with himself.

Jacobsen was looking uncertainly from one to the other of us. 'You mean the body that was in Willis Dexter's grave? That's Noah Harper?'

'The timing fits,' Gardner said. 'Except if Harper's dead, how did his fingerprints get to be on the weapon that killed Irving's dog?'

'Maybe the same way that Willis Dexter's came to be at the cabin,' Tom suggested.

There was a silence as we considered that. It had always been possible that Willis Dexter might not have faked his own death after all, that the killer had simply appropriated both his body and his fingerprints. But that couldn't have happened in this case.

'Were either of the hands missing from the corpse in Willis Dexter's casket?' Jacobsen asked.

'No,' I said. 'And all the fingers were there, too.'

'It's possible someone could've saved the film canister and steel bar with Dexter's and Harper's fingerprints already on them,' Tom suggested.

'The film canister, maybe. Dexter's print was smeared with a mineral oil that's used for most baby oils. There's no way of knowing how long it had been there,' Gardner said. 'But Harper's prints were left in the blood on the bar. It was only a few hours old.'

'Then the body from the casket can't be Noah Harper's. It's just not possible,' Jacobsen insisted.

Nobody said anything. Logic said she was right, not if the fingerprints had been left that morning. But judging from the expressions in the office no one felt very confident.

Tom took off his glasses and began to clean them. He looked more tired and somehow vulnerable without them. 'You might as well tell them what else you've found, David.'

Gardner and Jacobsen listened in silence as I described finding the pupal cases and dragonfly naiad in the casket, and the intact hyoid and pink teeth of the exhumed body.

'So it looks as though Terry Loomis and whoever was in the casket were killed the same way,' Gardner said when I'd finished. He turned to Tom. 'And you think these pink teeth could have been caused by strangulation?'

'Seems more likely than drowning,' Tom agreed mildly, and I tried not to smile. He hadn't mentioned Gardner's jibe at me in the cabin, but he obviously hadn't forgotten it. 'There wouldn't be much doubt at all if not for the obvious blood loss and wounds on Loomis's body.'

Gardner rubbed the back of his neck. 'The spatter patterns in the cabin looked authentic. But there's no way of knowing for sure if the blood came from Loomis until we get the DNA results.'

'That'll take weeks,' Tom commented.

'Tell me about it. It's times like this I wish we still did blood grouping. That'd at least tell us if the blood was the same type as his. But that's progress for you.' His expression made it clear what he thought of that. 'I'll get on to the lab. They're supposed to be fast-tracking this already, but I'll see if they can't speed things up a little.'

He didn't sound hopeful. While DNA provided a much more accurate method of matching and identification than the old technique of blood grouping, the testing process was also frustratingly slow. It was the same on both sides of the Atlantic; I'd heard more than one UK police officer complain that lab work took far longer than was portrayed on film or TV. The fact was that in the real world, fast-tracked or not, such things could take months.

Tom examined the lenses of his glasses, then resumed polishing them. 'You still haven't answered my question, Dan. Should we be worried?'

Gardner threw up his hands. 'What do you want me to say, Tom? I can't read this guy's mind; I don't know what he's going to do next. I wish I could. But even if he *is* responsible for Irving's disappearance it doesn't mean anyone else working on the case is in danger. I'm sorry as hell about Irving, but let's face it, the man courted publicity. Going on TV like that could have stirred up any number of psychos, not just this one.'

'Then we should just carry on like nothing's happened?'

'Within reason, yes. If I thought there was any real risk, believe me, I'd slap a twenty-four-hour guard on all of you. As it is, provided you take reasonable precautions, I'm sure there's no reason to worry.'

' "Reasonable precautions"?' Tom repeated impatiently. 'What's that mean? Don't take candy from strangers?'

'It means don't go walking dogs in woods by yourself,' Gardner retorted. 'Don't go down dark streets alone at night. C'mon, Tom, I don't have to spell it out.'

No, you don't. I thought about the scare the security guard had given me the night before. Perhaps I'd park somewhere less isolated in future.

'All right. Reasonable precautions it is,' Tom agreed, though he didn't sound happy. He put his glasses back on. 'So what do you think the chances are of finding Irving?'

'We're putting our full resources into it,' Gardner said, his guardedness returning.

Tom didn't press. We all knew exactly what Irving's chances were. 'Will you be bringing in another profiler?'

'That's under consideration,' Gardner said carefully. 'We haven't discounted Irving's profile of the killer altogether, but we're also looking at alternative viewpoints. And Diane's come up with an interesting theory.'

Colour bloomed on Jacobsen's otherwise impassive features. The blush reflex is a hard one to control. For someone who seemed to cultivate such outward composure, I imagined it must be infuriating.

'With all due respect to Professor Irving, I don't think the killings are sexual in nature, or that the killer is necessarily homosexual,' she said. 'I think Professor Irving might have become distracted by the fact that both victims were male and naked.'

She'd voiced the same views when the profiler had gone to see Terry Loomis's body in the cabin, and been put in her place for daring to disagree. For Irving's sake, I found myself hoping she was right.

'So how would you explain it?' Tom asked.

'I wouldn't, not yet. But the killer's actions suggest that he's not following a sexual agenda.' She was talking to Tom as an equal now, any reticence forgotten. 'We've got two crime scenes, and two sets of fingerprints from individuals who are very probably victims themselves. And then there're the hypodermic needles embedded in the body in Willis Dexter's grave, waiting for us to exhume it. The killer's showing off, running us round in circles to show who's in charge. It isn't enough for him to kill, he wants *recognition*. I'd agree with Professor Irving that the killings show evidence of pathological narcissism, but I'd say it goes further than that. This is more psychiatric territory than mine, but I think the killer bears all the hallmarks of a malignant narcissist.'

Tom looked blank. 'You'll have to excuse me, but I haven't a clue what that means.'

Jacobsen was too involved by now to be embarrassed. 'All narcissists are self-obsessed, but malignant narcissists are at the top of the scale. They have a pathological self-belief – a sense of grandiosity, even – which demands attention and admiration. They're convinced they're special in some way and want other people to acknowledge it as well. Crucially, they're also sadists who lack any conscience. They don't necessarily get fulfilment from inflicting pain, but they enjoy the sense of power it gives them. And they're indifferent to any suffering they might cause.'

'That sounds like a psychopath,' I said.

Jacobsen's grey eyes turned to me. 'Not quite, although there are shared characteristics. While a malignant narcissist is capable of extreme cruelty, he or she can still feel admiration and even respect for other people, provided the object of their respect displays what they consider "suitable" characteristics – generally a degree of success or power. According to Kernberg—'

'I don't think we need the footnotes, Diane,' Gardner told her.

Jacobsen looked chastened, but went on. 'The bottom line is I think we're dealing with someone who needs to demonstrate his superiority, maybe to himself as much as to us. He's got a chip on his shoulder and feels his talents and true worth aren't appreciated. That'd explain the lengths he's gone to, and also why he reacted as he did to what Professor Irving said on TV. He wouldn't only be infuriated at being publicly belittled, he'd hate to see someone else stealing his limelight.'

'Assuming this guy is also responsible for what happened to Irving,' Gardner put in, giving her a warning look.

'You sound like a damn lawyer, Dan,' Tom told him, but without heat. He gazed into space, absently tapping his chin with a finger. 'What about the employees from the funeral home? Do they all have alibis for when Irving went missing?'

'We're checking now, but to be frank I can't see any of them being behind this. The only two we've found so far who worked there around the time of Willis Dexter's funeral are both in their seventies.'

'What about York himself?'

'He claims to have been at work since five o'clock this morning. And before you ask, no there isn't anyone who can corroborate that,' Gardner said, with the air of someone backed into a corner.

'There's a surprise,' Tom muttered. 'Any sign of this mystery employee he claims he hired?'

'Dwight Chambers? We're still looking into it.'

'Meaning no.'

Gardner sighed. 'York's still a suspect. But whoever's behind this is too smart to bring all this attention down on himself. We're carrying out a full-scale search of Steeple Hill, and this time tomorrow the press are going to be all over the place. York's business is as good as dead no matter what happens.' He grimaced as he realized what he'd said. 'And the pun was unintentional.'

'From what I saw, it couldn't have carried on much longer anyway.' Light glinted on Tom's glasses as he stood up from behind the desk. 'Maybe York would rather go out with a bang.'

Or perhaps he's just another victim. But I kept that thought to myself.

It was growing dark as I pulled on to the quiet, tree-lined road where Tom and Mary lived. I would have worked late again if not for the dinner invitation, and after the day's interruptions I'd felt frustrated at having to break off. But not for long; as soon as I stepped out of the morgue into the sunny evening, I felt the iron fingers of tension release their hold on the back of my neck. I'd not really been aware of them until then, but Irving's disappearance, coming after what had happened to Kyle the day before, had shaken me more than I'd thought. Now the prospect of a few drinks and food with friends seemed like the perfect tonic.

The Liebermans' home was a lovely timber-framed house, white-painted and set well back from the road. It didn't seem to have changed from the first time I'd seen it, except for the majestic old oak that dominated the front lawn. On my last visit it had been in its prime; now it was in decline, and half of the sweeping branches were dead and bare.

Mary greeted me at the door, standing on tiptoe to kiss my cheek. 'David! Good of you to come.'

She had aged better than her husband. Her sandy hair had paled but retained its natural colour, and though her face was lined it still shone with health. Not many women in their sixties can wear jeans and get away with it, but Mary was one of them.

'Thank you, how lovely,' she said, taking the bottle of wine I'd brought. 'Come on through to the den. Sam and Paul aren't here yet, and Tom's on the phone with Robert.'

Robert was their only son. He worked in insurance and lived in New York. I'd never met him and Tom didn't talk about him much, but I had the impression that it wasn't an easy relationship.

'You're looking well,' Mary told me, leading me down the hall. 'Much better than you did last week.'

I'd had dinner with them on my first night. It already seemed a long time ago. 'Must be the sunshine,' I said.

'Well, whatever it is, it agrees with you.'

She opened the door into the den. It was actually an old conservatory, filled with healthy plants and cushioned rattan chairs. She settled me down in one with a beer, and then excused herself while she saw to dinner.

The panelled conservatory windows looked out over the back garden. I could just make out the tall shapes of trees in the darkness, outlined against the yellow lights of the next house. It was a nice neighbourhood. Tom had told me once that he and Mary had almost bankrupted themselves to buy the semi-derelict property back in the seventies, and never once regretted it.

I sipped the cold beer, feeling a little more tension slip away. Putting my head back, I thought about what had happened. It had been another broken day, with first the news about Irving and then Gardner and Jacobsen's visit taking me away from actual work.

Another distraction had come late that afternoon, with the arrival of the amino and volatile fatty acids analysis of Terry Loomis's tissue samples. Tom had come into the autopsy suite where I'd been processing the casket victim's remains.

'Well, we were wrong,' he'd declared without preamble. 'According to my calculations the time since death confirms the cabin manager's story. Loomis had only been dead for five days, not nearer seven like we thought. Here, see what you think.'

He handed me a sheet of figures. A quick look told me he was right, but Tom didn't make mistakes about things like that.

'Looks fine to me,' I said, returning them. 'But I still can't see how it can be.'

'Me neither.' He frowned down at the calculations as though offended by them. 'Even allowing for the heater being left on, I've never seen a body decompose to that extent after five days. There were pupating larvae on it, for God's sake!'

Blowfly larvae took six or seven days to pupate. Even if both Tom and I had been out in our time since death estimate, they shouldn't have reached that stage of their development for another day at least.

'Only one way they could have got there,' I said.

Tom smiled. 'You've been thinking it through as well. Go on.'

'Someone must have deliberately seeded the corpse with maggots.' It was the only thing that explained the condition of Terry Loomis's body. Fully grown larvae would have been able to get to work straight away, with no time lost waiting for the eggs to hatch. 'It wouldn't accelerate things by much, perhaps twelve to twenty-four hours at most. Still, with all the open wounds on the body it'd probably be enough.'

He nodded. 'Especially with the heater left on to raise the temperature. And there were way too many larvae on the body given that the cabin's doors and windows were all closed. Somebody obviously decided to give nature a boost. Clever, but it's hard to see what they hoped to gain, apart from muddying the water for a day or two.'

I'd been thinking about that as well. 'Perhaps that was enough. Remember what Diane Jacobsen said? Whoever's behind this is trying to prove something. Perhaps this was just another chance to show how clever he is.'

'Could be.' Tom gave me a thoughtful smile. 'Makes you wonder how he knows so much about it, though, doesn't it?' he said.

It had been a troubling thought.

I was still mulling that over when Tom came into the conservatory. He was freshly shaved and changed, with the deceptively healthy ruddiness that comes from a hot shower.

'Sorry about that. Our monthly duty call,' he said. The bitterness in his voice surprised me. He smiled, as though to acknowledge it, and lowered himself into a chair with a sigh. 'Has Mary fixed you up with a drink?'

I held up the beer. 'Yes, thanks.'

He nodded, but he still seemed distracted.

'Everything all right?' I asked.

'Sure.' He plucked irritably at the chair arm. 'It's just Robert. He was supposed to be visiting in a couple of weeks. Now it appears he won't have the time. I don't mind for myself so much, but Mary was looking forward to seeing him, and now . . . Ah, well. That's kids for you.'

The attempt to sound breezy faltered as he remembered my own circumstances. It was an innocent enough slip, but he looked relieved when the doorbell announced the arrival of Sam and Paul.

'Sorry we're late,' Paul said, as Mary ushered them into the conservatory. 'Got a flat tyre on my way home, and it took me ages to clean the damn oil off my hands.'

'You're here now. Samantha, you look positively radiant,' Tom said, going to kiss her. 'How are you?'

Sam lowered herself into a high-backed chair, made awkward by her swollen belly. With her blond hair pulled back into a ponytail, she looked fresh-faced and healthy. 'Impatient. If Junior doesn't hurry himself up we're going to have words before much longer.'

Tom laughed. 'You'll be doing the school run before you know it.'

His mood had lightened with their arrival, and by the time we sat down for dinner the atmosphere was easy and relaxed. Dinner was plain and unfussy – baked salmon with jacket potatoes and salad – but Mary was a good enough cook to make it seem special. As she served dessert, a hot peach pie with melting ice cream, Sam leaned across to me.

'How're you? You don't seem so tightly wound as last time I saw you,' she said, her voice low enough not to be overheard.

That had been in the restaurant where I thought I'd smelled Grace Strachan's perfume. It seemed like weeks ago, although it was only a few days. But a lot had happened since then.

'No, I don't suppose I am.' I smiled. 'I'm feeling pretty good, to be honest.'

She studied me for a moment or two. 'Yes, you look it.' Giving my arm a squeeze, she turned back to the main conversation.

After the meal, Mary and Sam disappeared into the kitchen to make coffee, rejecting our offers of help. 'You know as well as I do that you want to talk shop, and Sam and I have better things to discuss.'

'Anyone want to lay odds on it being babies?' Tom said after they'd gone out. He rubbed his hands. 'Well, I for one am going to have a bourbon. Care to join me? I have a bottle of Blanton's I need an excuse to open.'

'Just a small one,' Paul said.

'David? Or there's Scotch if you'd rather?'

'Bourbon's fine, thanks.'

Tom busied himself at a cabinet, taking out glasses and a distinctive bottle with a miniature horse and jockey perched on top. 'There's ice, but if I go into the kitchen Mary's going to read the riot act to me for drinking. And I'll take your disapproval as read, David.'

I hadn't been going to say anything. Sometimes abstinence can do more harm than good. Tom handed us each a glass, then raised his own.

'Your health, gentlemen.'

The bourbon was smooth with an aftertaste of burnt caramel. We sipped it, savouring it in silence. Tom cleared his throat.

'While you're both here there's something I wanted to tell you. It doesn't really concern you, David, but you might as well hear it as well.'

Paul and I glanced at each other. Tom stared pensively into his bourbon. 'You both know I was planning to bring my retirement forward to the end of summer. Well, I've decided not to wait that long.'

Paul set down his glass. 'You're joking.'

'It's time,' Tom said simply. 'I'm sorry to spring it on you like this, but . . . Well, it's no secret my health hasn't been good lately. And I have to think of what's fair to Mary. I thought the end of next month would be a good time. That's only a few weeks early, and it isn't as if the center will grind to a halt without me. I've got a feeling the next director should be a good one.'

That was aimed at Paul, but he didn't seem to notice. 'Have you told anyone else?'

'Only Mary. There's a faculty meeting next week. I thought I'd announce it then. But I wanted you to know first.'

Paul still looked stunned. 'Jesus, Tom. I don't know what to say.'

'How about "Happy retirement"?' Tom gave a smile. 'It isn't the end of the world. I'll still do some consultancy work, I dare say. Hell, I might even take up golf. So come on, no long faces. Let's have another toast.'

He reached for the bottle of Blanton's and topped up our drinks. There was a lump in my throat but I knew Tom didn't want us to be maudlin. I raised my glass.

'To fresh starts.'

He chinked his glass against mine. 'I'll drink to that.'

His announcement gave a bittersweet flavour to the rest of the evening. Mary beamed when she and Sam returned, but her eyes glittered with tears. Sam didn't try to hide hers, hugging Tom so hard he had to stoop over her pregnant stomach.

'Good for you,' she'd declared, wiping her eyes.

Tom himself had smiled broadly, and talked out his and Mary's plans, squeezing his wife's hand as he did so. But underlying it all was a sadness that no amount of celebration could disguise. This wasn't just a job Tom was retiring from.

It was the end of an era.

I was more glad than ever that I'd taken up his offer to help him on the investigation. He'd said it would be our last chance to work together, but I'd had no idea it was going to be the last time for him as well. I wondered if even he had, then.

As I drove back to my hotel just after midnight, I berated myself for not appreciating the opportunity I'd been given. Resolving to put any remaining doubts behind me, I told myself to make the most of working with Tom while it lasted. Another day or two and it would be all over.

At least, that's what I thought. I should have known better.

The next day they found another body.

The images form slowly, emerging like ghosts on the blank sheet of paper. The lamp casts a blood-red glow in the small chamber as you wait for the right moment, then lift the contact sheet from the tray of developing fluid and dip it into the stop bath before placing it in the fixer.

There. Perfect. Although you're not really aware of it, you whistle softly to yourself, a breathy, almost silent exhalation that holds no particular tune. Cramped as it is, you love being in the darkroom. It puts you in mind of a monk's cell: peaceful and meditative, a self-contained world in itself. Bathed in the room's transforming, carmine light, you feel cut off from everything, able to focus on coaxing to life the images implanted into the glossy photographic sheets.

Which is as it should be. The game you're playing, making the TBI and their so-called experts chase their own tails, might be a welcome relief and flattering to your ego. God knows, you deserve to indulge yourself after all the sacrifices you've made. But you shouldn't lose sight of the fact that it's only a diversion. The main thing, the real work, takes place in this small room.

There's nothing more important than this.

Getting to this stage has taken years, learning through trial and error. Your first camera was from a pawn shop, an old Kodak Instamatic that you'd been too inexperienced to know was poorly suited for your needs. It could capture the instant, but not in anything like enough detail. Too slow, too blurred, too unreliable. Not nearly enough precision, enough control, for what you wanted.

You've tried others since then. For a while you got excited about digital cameras, but for all their convenience the images lack – and here you smile to yourself – they lack the soul of film. Pixels don't have the depth, the resonance you're looking for. No matter how high the resolution, how true the colours, they're still only an impressionist approximation of their subject. Whereas film captures something of its essence, a transferral that goes beyond the chemical process. A real photograph is created by light, pure and simple: a paintbrush of photons that leaves its mark on the canvas of the film. There's a physical link between photographer and subject that calls for fine judgement, for skill. Too long in the chemical mix and the image is a dark ruin. Not long enough and it's a pallid might-have-been, culled before its time. Yes, film is undoubtedly more trouble, more demanding.

But nobody said a quest was supposed to be easy.

And that's what this is, a quest. Your own Holy Grail, except that you know for sure what you're searching for exists. You've seen it. And what you've seen once, you can see again.

You feel the usual nervousness as you lift the dripping contact sheet from the tray of fixer – carefully, having splashed fluid in your eyes once before –

and rinse it in cold water. This is the moment of truth. The man had been primed and ready by the time you got back, the fear and waiting bringing him to a hair-trigger alertness, as it always did. Though you try not to build up your hopes too much, you feel the inevitable anticipation as you scan the glossy sheet to see what you've got. But your excitement withers as you examine each of the miniature images, dismissing them one by one.

Blurred. No. No.

Useless!

In a sudden frenzy you rip the contact sheet in half and fling it aside. Lashing out at the developing trays, you knock them to the floor in a splash of chemicals. You raise your hand to swipe at the shelves full of bottles before you catch yourself. Fists knotted, you stand in the centre of the darkroom, chest heaving with the effort of restraint.

The stink of spilt developing fluids fills the small chamber. The sudden anger fades as you stare at the mess. Listlessly, you start to pick up some of the torn scraps, then abandon the effort. It can wait. The chemical fumes are overpowering, and some liquid splashed on to your bare arm. It's stinging already, and you know from past experience that it'll burn if you don't wash it off.

You're calmer as you leave the darkroom, the disappointment already shrinking. You're used to it by now, and there's no time to dwell on it. You have too much to do, too much to prepare. Thinking about that puts a spring back in your step. Failure's always frustrating, but you need to keep things in perspective.

There's always next time.

11

Tom called me before I left the hotel next morning. 'The TBI have found human remains at Steeple Hill.' He paused. 'These haven't been buried.'

Rather than take two cars he came to the hotel to pick me up. There was no debate this time over whether I would accompany him, only a tacit agreement that he wasn't going to try to manage by himself. I'd wondered what sort of mood he'd be in after the night before, whether there'd be any regret over announcing his retirement. If there was he hid it well.

'So . . . How are you feeling?' I asked, as we set off.

He hunched a shoulder in a shrug. 'Retirement won't be the end of the world. Life goes on, doesn't it?'

I agreed that it did.

The sun was out this time as we approached the paint-flaking gates to Steeple Hill. The thick pine woods bordering the lawns looked impenetrable, as though it were still night amongst their close-packed trunks.

Uniformed police officers stood outside the cemetery gate, barring entry to the press who had already assembled outside. Word that something had been found had obviously leaked. Coming on top of the exhumation, it had served as blood in the water to the news-hungry media. As Tom slowed down to show his ID, a photographer crouched to take a shot of us through the car window.

'Tell him he can have my autograph for ten dollars,' Tom grumbled, pulling inside.

We drove past the grave we'd exhumed last time and up to the main building. Steeple Hill's chapel looked to have been built in the 1960s, when American optimism had extended even into the funeral industry. It was a cheap attempt at modernism, a flat-roofed, single-storey block that aspired to Frank Lloyd Wright but fell woefully short. The coloured glass bricks that made up one wall beside the entrance were grimy and cracked, and the proportions were wrong in a way I couldn't quite put my finger on. A steeple was perched on top of the flat roof, looking as incongruous as a witch's hat on a table. Mounted on its peak was a metal cross that resembled two rusty girders badly welded together.

Gardner was standing outside the chapel, talking to a group of forensic agents, their white overalls grimed and filthy. He came over when he saw us.

'It's round the back,' he said without preamble.

A sudden sun-shower came from nowhere as we followed him round the side of the chapel, filling the air with silvered drops. It stopped as quickly as it started, leaving tiny rainbow prisms of light glistening on the grass and shrubs. Gardner led us down a thin gravel path that grew increasingly sparse and weed-choked the further we went. By the time we reached the tall yew hedge that screened the rear from view it was little more than a track worn in the grass.

But if the front of the chapel was run down, it was behind the hedge that Steeple Hill's true shabbiness was revealed. An ugly, utilitarian extension backed on to an enclosed yard that was strewn with rusting tools and empty containers. Squashed cigarette stubs littered the floor near the open back door like dirty white lozenges.

An air of neglect and dilapidation hung over it, and presiding over it all were the flies, weaving round in excited circles over the refuse.

'That's the mortuary in there,' Gardner said, nodding towards the extension. 'The crime scene team haven't found anything yet, but the Environmental Protection Agency aren't too happy about York's housekeeping.'

The sound of raised voices came to us as we neared the doorway. Inside I could see Jacobsen, a good head smaller than the three men she was with, but with her chin lifted defiantly. I guessed two of the men were the EPA officials Gardner had mentioned. The third was York. His voice was a near shout, trembling with emotion as he stabbed a finger in the air.

'. . . outrage! This is a respectable business! I will not be subjected to all sorts of *insinuations*—'

'No one's insinuating anything, sir,' Jacobsen cut in, politely but firmly. 'This is part of an ongoing homicide investigation, so it's in your own interests to cooperate.'

The funeral director's eyes were bulging. 'Are you *deaf*? I've already *told* you I don't know anything! Have you any idea of the *damage* this is doing my reputation?'

It was as though he didn't see the squalor around him. He broke off mid-tirade as he noticed us passing.

'Dr Lieberman!' he shouted, hurrying out towards us. 'Sir, I'd appreciate it if you'd help clear up this misunderstanding. As one professional to another, can you explain to these people that I have nothing to do with any of this?'

Tom took an involuntary step backwards as the funeral director bore down on him. Gardner moved in between them.

'Dr Lieberman's here on TBI business, Mr York. Go back inside and Agent Jacobsen will—'

'No, I will not! I am *not* going to stand by and see the good name of Steeple Hill dragged in the mud!' In the morning sunlight I could see that York's suit was grubby and creased, and a greasy scurf mark

145

striped his shirt collar. He hadn't shaved, and a frosting of grey whiskers crusted his jowls.

Jacobsen had come to flank him, so that between her and Gardner the funeral director had nowhere to go. Next to his seediness, she looked freshly minted. I caught a waft of soap and a clean, unfussy scent from her.

But there was no softness in her tone, and she held herself with a poised readiness. 'You need to come back inside, sir. The gentlemen from the Environmental Protection Agency still have questions to ask.'

York allowed her to steer him back towards the building, but continued to stare back at us over his shoulder.

'This is a conspiracy! A *conspiracy*! You think I don't know what's going on here? *Do you?*'

His voice echoed after us as Gardner ushered Tom away. 'Sorry about that.'

Tom smiled, but he looked shaken. 'He seems pretty upset.'

'Not as upset as he's going to be.'

Gardner led us towards the trees behind the chapel's mortuary. The funeral home backed on to a substantial pine wood. Crime scene tape had been strung between the trunks, and through the branches I glimpsed white-suited figures at work.

'One of the dogs found the remains in there,' Gardner said. 'They're pretty well scattered, but from a single individual so far's we can tell.'

'Definitely human?' Tom asked.

'Looks like. We weren't sure at first because they're so badly gnawed. Then we found a skull so it seems safe to assume they're a matching set. But after Tri-State we aren't taking any chances.'

I didn't blame him. The Tri-State Crematory in Georgia had made worldwide headlines back in 2002, when inspectors had found a human skull in its grounds. It proved to be the tip of a grisly iceberg. For no reason that was ever satisfactorily explained, the owner had simply kept many of the bodies he should have been cremating. Over three hundred human remains had been crammed into tiny vaults or stacked on top of each other in the surrounding forest. Some were even found dumped at the owner's house. Still, bad as Tri-State had been, there was one important difference from the current situation.

None of the victims there had been murdered.

Gardner took us over to the edge of the woods, where a trestle table stood laden with masks and protective gear. A few yards away, the pines formed an almost solid wall.

The TBI agent looked at Tom doubtfully, as though only now wondering about what he was asking of him. 'You sure you're OK to do this?'

'I've been in worse places.' Tom had already started opening a pack of disposable overalls. Gardner didn't seem convinced, but when he realized I was watching he erased the concern from his face.

'Then I'll let you get to it.'

I waited until he'd gone back to the mortuary. 'He's right, Tom. It's going to be uncomfortable in there.'

'I'll be fine.'

There was a stubbornness about him that told me I was wasting my time arguing. I zipped myself into the overalls and pulled on gloves and disposable overshoes. When Tom was ready we headed into the woods.

A hush enveloped us, as though the world outside had been abruptly cut off. Pine needles shivered all around, an eerie sound in the graveyard setting, like the whispering of the dead. A thick mat of them lay like coir matting underfoot, pebbled with fallen cones. The clean scent of pine that seeped through my mask was a welcome relief after the squalor of the funeral home.

But it was short-lived. The air was thick and still underneath the pines, untouched by any breeze. Almost immediately I felt myself begin to sweat as we stooped under the low branches and made our way towards the nearest white-clad agents.

'So what have you found?' Tom asked, trying to disguise his breathlessness as they made way for us.

It was hard to pick out individuals under the billowing protective gear and masks, but I recognized the big man who answered from the mountain cabin. *Lenny? No, Jerry.* His face was flushed and beaded with sweat above the mask, his overalls grimy with pine needles and bark.

'Oh, Lord, this is gonna be a day,' he panted, straightening. 'Got a skull and what's left of a ribcage, plus a few other bones. They're scattered pretty good, even the bigger ones. There's a fence further on back there, but it's too fallen down to stop anything getting in. On four legs or two. And these goddamn trees are a real bitch.'

'Any clothes?'

'Nope, but we got something that looks like an old sheet. Body could've been wrapped in that.'

Leaving him there, we made our way towards the nearest find. The forest floor was dotted with small flags, like an unkempt putting green, each marking a separate discovery. The one closest to us had been planted by what remained of a pelvis. It lay under a tree, so that we had to bend almost double to reach it, slipping on the friction-less carpet of pine needles. I glanced at Tom, hoping this wasn't going to be too much for him, but with the mask concealing much of his face it was hard to tell.

The pelvis was so badly chewed it was difficult to say whether it was male or female, but the femur lying next to it gave a better indication. Even though both ends of the big thigh bone were scored and pitted by animal teeth, it was obvious from its length that it was a man's.

'Quite a size,' Tom said, squatting down to examine it. 'How tall would you say its owner was?'

'Well over six feet. How tall was Willis Dexter?'

'Six two.' Tom smiled behind his mask, obviously thinking the same as me. It was starting to look as if we might have found the man who was supposed to have been buried at Steeple Hill. 'OK, let's see what else there is.'

Branches scratched at us, showering us with needles as we pushed through the trees. Tom was showing no obvious signs of discomfort, but it was heavy going. Sweat was running down my face, and I was beginning to cramp from being forced to walk in a permanent crouch. The pine scent was nauseating now, making my skin itch inside the constraining overalls.

The remains of what had once been a sheet lay some distance from the pelvis. Filthy and shredded, it had been marked with a different colour flag to distinguish it from the body parts. Near it, partially camouflaged by fallen pine needles, was a ribcage. A few ants scurried busily over it, foraging for any last vestiges of flesh, but there was little left. The bones had long since been picked clean, and the sternum and several smaller ribs were missing.

'Looks like this was where the body was dumped,' Tom commented, as I took photographs. 'The scattering looks pretty typical. Animals rather than dismemberment, I'd say.'

Nature abhors waste, and a body lying outdoors soon becomes a food source for the local wildlife. Dogs, foxes, birds and rodents – even bears in some parts of the US – will attend the feast, detaching and carrying away whatever they can. But because the bulkier torso is too big for all but the largest scavengers to move, it tends to be eaten *in situ*. That means the ribcage usually marks the location where the body originally lay.

Tom peered at the end of one of the ribs. He beckoned me closer. 'See here? Saw marks.'

Like most of the other bones, the rib had been badly gnawed. But parallel lines were still visible among the teeth marks, fine striations running across the bone's end.

'Hacksaw blade, by the look of it. The same as you'd get from an autopsy,' I said. Standard procedure during an autopsy was to cut the ribcage on either side of the sternum, so that it could be removed to give access to the organs underneath. Bone cutters were sometimes used, but an electric saw was often faster.

That would have produced marks just like these.

'Starting to look more and more like we've found Willis Dexter, isn't it?' Tom said. He started to push himself to his feet. 'Male, right height, with autopsy cuts on his ribs. And Dexter's clothes were burned in the car crash. Without any family to provide more, chances are the body would be left in the sheet it came in from the morgue. Time scale's about right, too. There's no moss or lichen on the bones, so they've been here less than a year. That seems—'

He gave a sudden gasp and doubled up, clutching at his chest. I pulled off his mask and had to hide my alarm when I saw the waxy pallor of his face.

'Where are your tablets?'

His mouth was stretched in a grimace. 'Side pocket . . .'

I tore open his overalls, berating myself. *You should never have let him do this!* If he collapsed in here . . . There was a button-down pocket on the thigh of his chinos. I pulled it open but couldn't find any tablets.

'They're not there.' I tried to sound calm.

His eyes were screwed shut with pain. His lips had developed a blue tinge. 'Shirt . . .'

I patted his shirt pocket and felt a squat hard shape. *Thank God!* I pulled it out and unscrewed the top, shaking out one of the tiny pills. Tom's hand trembled as he slipped it under his tongue. Nothing happened for a few moments, then the tightness in his face began to relax.

'OK?' I asked. He nodded, too drained to speak. 'Just take it easy for a minute or two.'

There was a rustle from nearby as Jerry, the big forensic agent, came over. 'Y'all OK?'

I felt Tom's hand tighten on my arm before I could answer. 'Fine. Just need to catch my breath.'

The agent didn't look fooled, but left us alone. As soon as he'd gone Tom's shoulders slumped again.

'Can you walk?' I asked.

He drew in an unsteady breath. 'I think so.'

'Come on, let's get you out of here.'

'I'll manage. You carry on.'

'I'm not letting you—'

He gripped my arm again. There was a quiet entreaty in his eyes. 'Please, David.'

I didn't like the idea of letting him make his way from the woods by himself, but it would only agitate him more if I insisted on going as well. I looked between the pine trunks to the edge of the trees, gauging how far it was.

'I'll take it nice and slow,' he said, guessing what I was thinking. 'And I promise to rest as soon as I get out.'

'You need to see a doctor.'

'I just have.' He gave a weak smile. 'Don't worry. You just finish off here.'

Anxiously, I watched as he picked his way through the woods, moving with the deliberation of an old man. I waited until he'd reached the tree line, vanishing through the close-pressed branches into the daylight before I went over to where Jerry was examining an object on the ground that might or might not have been a piece of bone. He glanced up as I approached.

'He all right?'

'Just the heat. You said earlier that you'd found a skull?' I went on quickly.

He led me to where another small flag had been set at the bottom of a slope. The pale dome of a human cranium sat next to it, half buried among the pine needles. The mandible was missing, and the skull lay upside down like a dirty ivory bowl. The heaviness of its structure suggested it was a man's, and I could make out fracture lines radiating across the frontal bone of the forehead. It was

151

the sort of injury caused by impact with something flat and hard.

Like a car windscreen.

I was sure now that the remains belonged to Willis Dexter, in which case we probably wouldn't learn much from them. It was almost certain that the mechanic had died in a car crash rather than been murdered. His only connection with the killings was that his casket and grave had been appropriated by the killer. If we could have established if either of his hands, or even any digits, were missing it might at least explain how his fingerprints came to be left on the film canister so long after his death. But no carpels or phalanges had been found, and given the size of the woods it was unlikely that they ever would be. The remains had been too thoroughly picked over by scavengers. Even if the smaller bones hadn't been eaten, they could be anywhere by now.

'Wasted journey, huh, doc?' Jerry said cheerfully as I photographed the latest find – a rib chewed down to half its original size. 'Not much to say, other than they're human. And we could've told you that. Anyhow, if you're done we'd like to start getting this all boxed and bagged.'

It was an unsubtle hint. I was about to leave him to it when I noticed another flag.

'What's over there?'

'Just some teeth. Must've come loose when the jaw was pulled off.'

There was nothing unusual about that. Scavengers generally eat the face first, and the teeth could easily have been dislodged from the missing mandible. I almost didn't bother going over. I was hot and tired, and wanted to see how Tom was. But I'd learned from hard experience not to take anything for granted.

'I'd better take a look,' I said.

The flag had been placed amongst the exposed roots of a scrubby pine. It wasn't far from where the ribcage lay, but it wasn't until I was up close that I could make out the dirty nuggets of ivory. There were four molars, coated in dirt and hard to see amongst the pine needles.

It was a testament to the thoroughness of the search that they'd been found at all. Yet as I looked at them it seemed that something wasn't quite right . . .

The heat and discomfort were instantly forgotten as I realized what it was.

'Just teeth, like I told you. So, you done now?' Jerry asked as I began to photograph them. The hint was plainer this time.

'Have you got photos of these yourself?'

He gave me a look that said I was an idiot for asking. 'Doc, we've got photographs coming out of the wazoo.'

I pushed myself to my feet. 'I'd take some more of these anyway. You're going to need them.'

Leaving him staring after me I made my way out of the woods. Sweat was trickling down my back as I left the claustrophobic cover of the pines and gratefully pulled off my mask. Unfastening my overalls, I ducked under the crime scene tape and looked around for Tom. He was standing some way off, talking to Gardner and Jacobsen in the shade of the yew hedge. He looked OK, but my relief lasted only until I saw Hicks was with them. A moment later I heard the raised voices.

'. . . no legal standing in this investigation! You know that as well as I do.'

'That's ridiculous. You're just splitting hairs, Donald,' Tom said.

'Splitting *hairs*?' The sun glinted off the pathologist's bald head as he thrust out his chin. 'Will the judge be "splitting hairs" when he throws out a homicide case because an expert witness let an unsupervised *assistant* tramp all over a crime scene? One who probably won't even be in the *country* when this goes to court?'

It wasn't hard to guess who they were talking about. They all fell silent as I approached.

'How are you feeling?' I asked Tom. First things first.

'I'm fine. I just needed some water.'

Up close I could see he was still pale, but he seemed a lot better

153

than he had. The look he gave me made it clear I shouldn't mention his attack in front of the others.

I turned to Gardner. 'Is there a problem?'

'You're damn right there's a problem!' Hicks interrupted. For all his indignation, I could see he was enjoying himself.

'Maybe we should discuss this some other time,' Gardner suggested wearily.

But the pathologist wasn't going to be diverted. 'No, this needs to be settled now. This is one of the biggest serial killer investigations the state's seen in years. We can't risk amateurs fouling things up.'

Amateurs? I clamped my mouth shut as my temper threatened to slip. Whatever I said would only make things worse.

'David's every bit as competent as I am,' Tom said, but he lacked the energy to argue. Hicks stabbed a finger at him.

'That's irrelevant! He shouldn't have been wandering around a crime scene by himself. What about it, Gardner? You going to start handing out *tickets* so anyone can just walk in?'

Gardner's jaw muscles knotted, but that had hit home. 'He's got a point, Tom.'

'Goddammit, Dan, David's been doing us a *favour*!'

But I'd heard enough. It was obvious where this was going. 'It's all right. I don't want to make things difficult.'

Tom looked stricken, but Hicks was barely able to suppress his glee.

'No offence, Dr . . . Hunter, is it? I'm sure you're well enough respected back home, but this is Tennessee. This isn't your affair.'

I didn't trust myself to say anything. Jacobsen was staring at Hicks with an unreadable expression. Gardner looked as though he wished the whole thing was over with.

'I'm sorry, David,' Tom said helplessly.

'It's OK.' I handed him the camera. I just wanted to be somewhere else. Anywhere. 'Will you be able to manage?'

I didn't want to say more, not in front of the others, but Tom knew

154

what I meant. He gave a quick, embarrassed nod. I started to turn away before I remembered what I needed to tell him.

'You should take a look at the teeth that've been found in there. They don't belong with the rest of the remains.'

'How do you know?' Hicks demanded.

'Because they're from a pig.'

That silenced him. I saw the flash of interest in Tom's eyes. 'Premolars?'

I nodded, knowing he'd understand. But he was the only one. Hicks was glaring at me as though he suspected some sort of trick.

'You're telling me they've found *pig's* teeth? What the hell are they doing there?'

'Don't ask me. I'm only an amateur,' I said.

It was a cheap parting shot, but I couldn't help myself. As I walked away I saw the smile on Tom's face, and thought there might even have been a ghost of one on Jacobsen's.

But it didn't make me feel any better. I retraced my steps round to the front of the chapel, yanking the overalls' zip so hard they tore. I wrenched myself free and stuffed them in a plastic bin already half full of discarded protective gear. When I stripped off the rubber gloves sweat dripped out of them, forming dark splashes like a modernist painting in the dirt. My hands were pale and wrinkled from being trapped in the airless latex, and for an instant I felt a tug of something like déjà vu.

What? What does that remind me of?

But I was too angry to dwell on it. And a more mundane thought had occurred to me. I'd come to Steeple Hill in Tom's car. After my grand exit, now I was stranded out here.

Oh, terrific. I flung the gloves into the bin and took out my phone before realizing I didn't know the numbers of any local taxis. And even if I did, they wouldn't be allowed into the cemetery.

I swore under my breath. I could always wait for Tom to finish, but my pride wouldn't allow that. *Fine. I'll walk.* Knowing I was being

155

stubborn but in too foul a temper to care, I headed for the gates.

'Dr Hunter!'

I turned to see Jacobsen coming along the path towards me. The bright sun was in her face, making her squint slightly against the glare. It caused tiny crow's feet to appear at the corners of the grey eyes, giving her a quizzical, almost humorous look that softened her features.

'Dr Lieberman said you didn't have your car. How are you getting back to town?'

'I'll manage.'

'I'll drive you.'

'No thanks.' I was in no mood to accept favours.

Her expression was impossible to read as she brushed a wayward strand of hair from her face, tucking it neatly behind her ear. 'I wouldn't recommend walking. Not with all the press outside.'

I'd not thought about that. The anger began to leak away, leaving me feeling more than a little stupid.

'I'll get my car,' Jacobsen said.

12

The silence in the car wasn't exactly companionable, but neither was it awkward. I didn't feel talkative and Jacobsen didn't seem concerned either way. My temper had cooled a little, but there was still a slow burn of resentment that refused to die down.

I pulled at my shirt, still hot and uncomfortable from the time spent in the pine woods. The inside of the car had been turned into an oven by the sun, but the air conditioning was finally starting to win the battle. I stared moodily out of the window, watching the unending succession of stores and fast food restaurants troop past: glass, brick and concrete set against the dark green backdrop of the mountains. More than ever I was aware of how unfamiliar much of it was. I didn't belong here. *And you're certainly not wanted.*

Perhaps I should check for earlier flights after all.

'You might not like it, but Dr Hicks had a point,' Jacobsen said, rousing me from my thoughts. 'Dr Lieberman's an authorized TBI consultant. You aren't.'

'I know how to work crime scenes,' I said, stung.

'I'm sure you do, but this isn't about how capable you are. If this goes to trial we can't afford to have a defence attorney argue that we didn't follow procedure.' She turned to look at me, her grey eyes candid. 'You should know that.'

I felt my self-righteous anger wilt. She was right. And there was more at stake here than my pride.

'Dr Lieberman's ill, isn't he?'

The question took me by surprise. 'What makes you say that?'

Jacobsen kept her attention on the road. 'My dad had a bad heart. He looked the same way.'

'What happened?' I asked.

'He died.'

'I'm sorry.'

'It was years ago,' she said, closing the subject.

Her face was studiedly expressionless, but I sensed she was regretting giving away even that much about herself. It struck me again how attractive she was. I'd been aware of it before, of course, but only in an academic way, as you might admire the shape and form of a marble statue.

Now, though, in the close confines of the car, I was all too conscious of it. She'd taken off her jacket, and her short-sleeved white shirt showed off the toned muscles of her arms. Her gun was still clipped to her belt, a jarring note against the smart business suit. But I could hear the whisper of her skirt on her legs as she worked the pedals, smell the fresh clean scent from her skin; a scented soap, I guessed, too light to be perfume.

My sudden awareness of her was unnerving. I looked away from the full lips and stared resolutely ahead, keeping my eyes fixed on the road. Jacobsen would probably break my wrist if she realized what I was thinking. *Or shoot you.*

'Any news about Irving?' I asked, to take my mind off it.

'We're still searching.' *No, in other words.* 'Dr Lieberman says the remains in the woods were probably Willis Dexter's,' she said, businesslike again.

'It seems that way.' I described the fractures to the skull's forehead, and how they fitted Dexter's injuries. 'Makes sense, I suppose. Someone switched bodies, and then dumped Dexter's in the woods at the back, where it wouldn't be found unless the grounds were searched.'

'But whoever did that would know that would happen as soon as we found the wrong body in the grave. So they obviously wanted us to find this as well.'

First Loomis, then the unidentified remains in the casket, now Dexter. It was like a paper trail of corpses, each one leading to the next. 'It had to be someone with access to Steeple Hill,' I said. 'Have you got any further in tracking down this Dwight Chambers who York claims was working there?'

'We're still looking into it.' Jacobsen slowed the car to a stop as we drew up to a red traffic light. 'You sure the teeth you saw were from a pig?'

'Certain.'

'And you think they were left deliberately?'

'There's no other reason for them to be there. They were above the ribcage, exactly where the head would have been before scavengers got to the body. But none of the teeth showed any signs of scoring or damage, and if there'd been any gum tissue on them rodents would have gnawed it off. Which suggests the teeth were already clean when they were left there.'

There was a small furrow between Jacobsen's eyes. 'But what's the point?'

'Don't ask me. Perhaps whoever left them there just wanted to show off again.'

'I don't follow. How is leaving pig's teeth showing off?'

'Pig premolars look a lot like human molars. Unless you know what you're looking for, it's easy to mistake one for the other.'

Jacobsen's frown lifted. 'So the killer was letting us see he knows about details like that. Like the fingerprints left at the crime scenes. He's not just testing us, he's bragging how clever he is.'

159

She gave a start as a horn blared behind us, alerting us that the lights were green. Flustered, she pulled away. I looked out of the window so she wouldn't see my smile.

'It sounds like pretty specialist knowledge. Who'd have access to that sort of information?' she went on, her composure once more in place.

'It's no secret. Anyone with—'

I stopped short.

'With a forensic background?' Jacobsen finished for me.

'Yes,' I admitted.

'Such as forensic anthropology?'

'Or forensic archaeology, or pathology. Or any one of a dozen different forensic disciplines. Anyone who can be bothered to look through textbooks can find that sort of information. It doesn't mean you have to start pointing fingers at people who work in the field.'

'I wasn't pointing fingers at anyone.'

The silence that fell now was anything but comfortable. I searched for a way to break it, but the aura around Jacobsen made small talk unthinkable. I stared out of the window, feeling flat and tired. Traffic streamed past, glinting in the early afternoon sunshine.

'You don't think much of psychology, do you?' she said suddenly.

I wished I hadn't said anything, but there was no avoiding it now. 'I think there's too much reliance on it sometimes. It's a useful tool but it isn't infallible. Irving's profile showed that.'

Her chin came up. 'Professor Irving let himself be sidetracked by the fact that both victims were male and naked.'

'You don't think that's significant?'

'Not that they're male, no. And I think you and Dr Lieberman hit on the reason why they were naked.'

That threw me, but only for a second. 'A naked body decomposes faster than one with clothes on,' I said, annoyed with myself for not having seen it sooner.

She gave a nod. She seemed as keen to skirt past the brief

awkwardness as I was. 'And both Terry Loomis's body and the exhumed remains were more decomposed than they'd any right to be. It isn't unreasonable to assume they were both unclothed for similar reasons.'

Another chance for the killer to sow confusion and demonstrate his cleverness. 'The exhumed body would have to have been stripped for the needles to be planted anyway,' I said. 'And once they were in place it'd be too risky to handle it any more than necessary. Certainly not just to put its clothes back on. But that doesn't alter the fact that all the victims were male.'

'The ones we know about, you mean.'

'You think there are more we haven't found yet?'

I thought at first I'd gone too far. Jacobsen didn't answer, and I reminded myself that she didn't have to; I was no longer a part of the investigation. *Get used to it. You're just a tourist now.*

But just as I was about to withdraw the question she seemed to reach a decision. 'This is pure speculation. But I'd agree with Professor Irving that we've only found the victims the killer wanted us to find. The level of brutality and sheer confidence he's displayed makes it almost certain that there are others. No one develops that sort of . . . *sophistication*, for want of a better word, first time round.'

That hadn't occurred to me before. It was a disturbing thought.

Jacobsen pulled down the visor as a curve in the road threw the sun in her face. 'Whatever the killer's agenda is, I don't think his victims' physical characteristics play a part in it,' she went on. 'We've got a thirty-six-year-old white insurance clerk, a black male in his fifties, and – in all probability – a forty-four-year-old psychologist, with no apparent connection between any of them. That suggests we're dealing with an opportunist who preys on random victims. Male or female, I doubt it makes any difference to him.'

'What about Irving? He wasn't random, he was deliberately targeted.'

'Professor Irving was an exception. I don't think he figured in the killer's plans until he went on TV, but when he did the killer acted straight away. Which tells us something important.'

'You mean apart from that he's a dangerous lunatic?'

A quick smile softened her features. 'Apart from that. Everything we have so far says that this is someone who deliberates and plans his actions carefully. The needles were planted in the body *six months* before he left Dexter's fingerprints at the cabin. That shows a methodical, ordered mind. But what happened with Professor Irving shows there's also another side. One that's impulsive and unstable. Prick his ego and he can't help himself.'

I noticed she wasn't even trying to pretend any more that Irving might not be another victim. 'Is that good or bad?'

'Both. It means he's unpredictable, which makes him even more dangerous. But if he acts on impulse then sooner or later he'll make a mistake.' Jacobsen squinted again as the sun reflected off the cars in front. 'My sunglasses are in my jacket. Could you pass them, please?'

The jacket was neatly folded on the back seat. I twisted round and reached for it. A waft of delicate scent came from the soft fabric, and I felt an odd intimacy as I searched its pockets. I found a pair of aviator shades and handed them to her. Our fingers brushed as she took them; her skin was cool and dry, but with an underlying heat.

'Thanks,' she said, putting on the sunglasses.

'You mentioned his agenda a moment ago,' I said quickly. 'I thought you'd already said that he craves recognition, that he's a . . . what was it? A "malignant narcissist"? Doesn't that explain it?'

Jacobsen inclined her head slightly. With her eyes concealed, she looked more unreadable than ever. 'It explains the extreme lengths he's prepared to go to, but not *why* he kills in the first place. He's got to get something out of it, have some pathological itch he's trying to scratch. If it isn't sexual, then what?'

'Perhaps he just enjoys inflicting pain,' I suggested.

She shook her head. The small v was visible again above the sunglasses. 'No. He might enjoy the sense of power it gives him, but it's more than that. Something's driving him to do all this. We just don't know yet what it is.'

The sunlight was abruptly blotted out as a black pick-up truck drew up alongside. It towered over the car, a petrol-guzzling monstrosity with tinted windows, then quickly pulled ahead. It had only just cleared us when suddenly it cut into our lane. My foot stamped reflexively on to the floor as I braced for a collision. But with barely a touch on the brake, Jacobsen swerved into the other lane, as smoothly as though the move were choreographed.

It was a cool display of driving, all the more impressive because she appeared unaware of it. She flicked an irritated glance at the pick-up as it accelerated away, but otherwise dismissed it.

The incident broke the mood, though. She grew distant again after that, either preoccupied with what we'd said or regretting saying as much as she had. In any event there wasn't any more time for conversation. We were already approaching the centre of Knoxville. My spirits sank further the closer we got. Jacobsen dropped me back at my hotel, her reserve now as unassailable as any wall. Her sunglasses hid her eyes as she drove off with the briefest of nods, leaving me on the pavement, stiff-muscled from hunching over in the pine woods.

I felt at a complete loss as to what to do next. I didn't know if my exclusion extended to the morgue, and didn't want to phone Tom to ask. Nor did I feel like going out to the facility, not until I'd a better idea of how things stood.

Standing there in the bright spring sunshine, with people bustling around me, the full extent of what had happened finally sank in. While I'd been with Jacobsen I'd been able to keep it at arm's length, but now I had to face up to it.

For the first time in my career I'd been thrown off an investigation.

I showered and changed, then bought a sandwich and ate lunch at the side of the river, watching the tourist-carrying paddleboats churn past. There's something about water that's primordially soothing. It

163

seems to touch some deep chord in our subconscious; stir some gene memory of the womb. I breathed in the faintly swampy air, watching a flight of geese heading upriver, and tried to tell myself that I wasn't bored. Objectively, I knew I shouldn't take what had happened at the cemetery personally. I'd been caught in Hicks's crossfire, collateral damage of professional politics that didn't concern me. I told myself that I shouldn't regard it as a loss of face.

It didn't make me feel any better.

After lunch, I wandered aimlessly around the streets, waiting for my phone to ring. It was a long time since I'd been in Knoxville, and the city had changed. The trolley cars were still there, though, and the golden mirror-ball of the Sunsphere remained an unmistakable feature on the skyline.

But I wasn't in the mood for sightseeing. My phone remained stubbornly silent, a dead weight in my pocket. I was tempted to call Tom, but I knew there was no point. He'd ring me when he could.

It was late afternoon when I finally heard from him. He sounded tired as he apologized for what had happened that morning.

'It's just Hicks trying to stir up a fuss. I'm going to talk to Dan again tomorrow. Once the dust has settled I'm sure he'll see sense. There's no reason why you can't carry on working with me at the morgue, at least.'

'What are you going to do in the meantime?' I asked. 'You can't manage by yourself. Why don't you let Paul help?'

'Paul's out of town today. But I'm sure Summer will lend a hand again.'

'You need to take it easy. Have you seen a doctor yet?'

'Don't worry,' he said, in a tone of voice that told me I was wasting my breath. 'I'm really sorry about this, David, but I'll sort it out. Just sit tight for now.'

There wasn't much else I could do. I resolved to try to enjoy the rest of the evening. *A little leisure time won't kill you.* The bars and cafés had started filling up, office workers stopping off on their way home.

The murmur of laughter and conversation was inviting, and on impulse I stopped at a bar with a wooden terrace overlooking the river. I found a table by the railing and ordered a beer. Enjoying the last of the afternoon sun, I watched the slow-moving Tennessee slide by, invisible currents forming dimples and swirls on the gelid surface.

Gradually, I felt myself begin to relax. By the time I'd finished my beer I couldn't see any pressing reason to leave, so I asked for the menu. I ordered a plate of seafood linguine and a glass of Californian Zinfandel. Just the one, I vowed, telling myself I should make an early start next day, regardless of whether I was helping Tom or not. But by the time I'd finished the rich, garlic-infused food, that no longer seemed quite such a compelling argument.

I ordered another glass of wine. The sun sank behind the trees, but it was still warm even as dusk began to settle. The electric lights that lit the terrace drew the first of the evening's moths. They bumped and whirred against the glass, black silhouettes against the white globes. I tried to recall visiting this stretch of river when I'd first come to Knoxville all those years ago. I supposed I must have at some point, but I'd no recollection of it. I'd rented a cramped basement apartment in a different – and cheaper – part of town, on the fringes of the increasingly gentrified old quarter. When I'd gone out I'd tended to go to the bars round there rather than the more expensive ones on the riverfront.

Thinking about that shook loose other memories. Out of nowhere the face of a girl I'd seen for a while came back to me. Beth, a nurse at the hospital. I hadn't thought of her in years. I smiled, wondering where she was now, what she was doing. If she ever thought about the British forensic student she'd once known.

I'd returned to England not long after that. And a few weeks later I'd met my wife, Kara. The thought of her and our daughter brought with it the usual vertiginous dip, but I was used enough to it by now not to be sucked in.

I picked up my mobile from the table and opened my list of contacts. Jenny's name and number seemed to jump out at me even before I'd highlighted them on the illuminated display. I scrolled through the options until I came to *Delete*, and held my thumb poised over the button. Then, without pressing it, I snapped the phone shut and put it away.

I finished the last of my wine and pulled my thoughts from the track they'd been following. An image of Jacobsen sitting in the car earlier replaced them, bare arms toned and tanned in the short-sleeved white top. It occurred to me that I didn't know anything about her. Not how old she was, where she was from or where she lived.

But I'd noticed there was no wedding ring on her left hand.

Oh, give it a rest. Still, I couldn't help but smile as I ordered another glass of wine.

It's darkening outside. Your favourite time. The point of transition between two extremes: day and night. Heaven and hell. The earth's rotation caught on the cusp, neither one thing nor the other, yet full of the potential of both.

If only everything were so simple.

You brush the camera lens carefully, then gently wipe it with a square of buttery soft chamois until the finely ground glass is mirror bright. Tilting the lens to catch the light, you examine it for any last speck of dust that might mar its perfect surface. There's nothing, but you polish it again anyway, just to be sure.

The camera is your most prized possession. The old Leica has seen some heavy use in the years since you bought it, and never once let you down. Its black and white images are always crystal clear, so sharp and fine-grained you could fall into them.

It isn't the camera's fault you haven't found what you're looking for.

You try to tell yourself that tonight will be just like all the other times, but you know it isn't. You've always operated under cover of obscurity before, been able to act with impunity because no one knew you existed. Now that's all changed. And even though it was your own decision, your own choice to emerge into the limelight, it alters everything.

For good or bad, you're committed now. There's no going back.

True, you've prepared for it. You wouldn't have started this without an exit strategy. When the time comes you'll be able to slide back into the shadows, just like before. But you've got to see it through to the end first. And while the rewards might be great, so is the risk.

You can't afford any mistakes.

You do your best to believe that what happens tonight doesn't matter in the greater scheme of things, that your real work will continue regardless. But it rings false. The truth is there's more at stake now. Although you hate to admit it, all the failures have taken their toll. You need this, you need the affirmation that you haven't wasted all these years.

Your entire life.

You finish polishing the camera lens and pour yourself a glass of milk. You ought to have something to soak up the acid in your stomach, but it's too knotted to eat. The milk's been opened for a day or two now, and the scum on top says it's probably turned. But that's one of the benefits of not being able to smell or taste anything. You drink it straight off, staring out of the window at the trees silhouetted against the sky. When you set the empty glass back on the kitchen table, the smeared interior gives it a ghostly translucency in the gathering dark.

You like that idea: a ghost glass.

But the pleasure soon fades. This is the part you hate most, the waiting. Still, it won't be much longer now. You look across the room at where the uniform hangs on the back of the door, barely visible in the deepening shadows. It wouldn't stand close inspection, but most people don't look too closely. They see only a uniform in those first few seconds.

And that's all you need.

You pour yourself another glass of milk, then stare out of the dirty window as the last of the light vanishes from the sky.

13

The dentist lay exactly as he had the last time I'd seen him. He was still sprawled on his back, lying with the immobility only the dead can achieve. But he'd changed in other ways. The flesh had dried in the sun, skin and hair slipping from him like an unwanted coat. After a few more days stubborn tendons would be all that remained of the soft tissue, and before much longer there would be nothing left but enduring bone.

I'd woken with a nagging headache, regretting the last glass of wine I'd had the previous night. Remembering what had happened before that hadn't made me feel any better. As I'd showered I'd wondered what I should do until I heard from Tom. But there was really no decision.

I'd had enough of being a tourist.

The car park had been nearly empty when I'd arrived at the facility. It was still in shadow, and I shivered in the early morning chill as I pulled on a pair of overalls. I took out my phone, weighing up whether or not to leave it on. Normally I turned it off before I went through the gates – there seemed something disrespectful about disturbing the quiet inside with phone conversations – but I didn't want to miss Tom's call. I was tempted to leave it on vibrate, except that then I'd spend all morning waiting for its telltale buzz. Besides, realistically I knew Tom wouldn't ring Gardner until later anyway.

Making up my mind, I switched off the phone and thrust it away.

Hoisting my bag on to my shoulder, I headed for the gates. Early as it was, I wasn't the first there. Inside, a young man and woman in surgical scrubs, graduate students by the look of them, were chatting as they made their way back down through the trees. They gave me a friendly 'Hi' as they passed, then disappeared about their business.

Once they'd gone, silence descended on the wooded enclosure. Apart from the birdsong, I might have been the only living thing there. It was cool inside, the sun not yet high enough to break through the trees. Dew darkened the bottoms of my overalls as I went up the wooded hillside to where the dentist's body lay. The protective mesh cage meant that, among other things, I could observe how his body decomposed when no insects or scavengers were able to reach it. It wasn't exactly original research but I'd never carried it out before myself. And charting something first hand was always better than relying on the work of others.

It had been a few days since I'd been here, though, so I'd some catching up to do. Stepping through a small door in the cage, I took a tape measure, calipers, camera and notepad from my bag and squatted down to work. I made heavy going of it; the headache was still a nagging throb behind my eyes, and the thought of the phone in my bag was a constant drag on my attention. When I found myself taking the same measurement twice I shook myself angrily. *Come on, Hunter, focus. This is what you came here for.*

Closing my mind to distractions, I buckled down to the task. Headache and phone were temporarily forgotten as I was drawn into the microcosm of decay. Viewed dispassionately, our physical dissolution is no different from any other natural cycle. And, like any other natural process, it has to be studied before it can be fully understood.

Eventually, sensations of discomfort began to make themselves known. My neck was stiff, and when I paused to flex it I realized I was hot and cramped. The sun was high enough now to reach through the trees, and I could feel myself starting to sweat in the overalls. Checking the time, I saw with surprise it was almost midday.

I stepped out of the cage and closed the door behind me, then stretched, wincing as my shoulder popped. Pulling off my gloves, I started to take a bottle of water from my bag, but stopped when I caught sight of my hands. The skin was pale and wrinkled after being in the tight rubber gloves. There was nothing unusual about that, yet for some reason the sight prompted something to bump against my subconscious.

It was the same sense of almost-recognition as I'd had the day before at Steeple Hill, and just as elusive. Knowing better than to force it, I took a drink of water. As I put the bottle away I wondered if Tom had spoken to Gardner yet. The temptation to switch on my phone to check for messages lured me for a moment, but I firmly put it aside. *Don't get distracted. Finish what you're doing here first.*

It was easier said than done. I knew there was a good chance that Tom would have called by now, and the awareness nagged at my concentration. Refusing to give in to it, I took almost perverse care over the last few measurements, checking and noting them down in a log book before I packed away. Locking the mesh cage behind me, I headed for the gates. When I reached my car I stripped out of my overalls and gloves and put everything in the boot before I allowed myself to turn on the phone.

It beeped straight away to let me know I had a message. I felt my stomach knot with anticipation. It had been left not long after I'd arrived at the facility, and I felt a stab of frustration when I realized I'd missed Tom's call by minutes.

But the message wasn't from him. It was from Paul, to tell me that Tom had had a heart attack.

★

We don't realize how reliant we are on context. We define people by how we normally see them, but take them out of that, place them in a different setting and situation, and our mind baulks. What was once familiar becomes something strange and unsettling.

I wouldn't have recognized Tom.

An oxygen tube snaked up his nose, and a drip fed into his arm, held in place by strips of tape. Wires ran from him to a monitor, where wavering electronic lines silently traced the progress of his heart. In the loose hospital gown, his upper arms were pale and scrawny, with the wasted muscles of an old man.

But then it was an old man's head that lay on the pillow, grey-skinned and sunken-cheeked.

The heart attack had struck at the morgue the night before. He'd been working late, wanting to make up for the time lost out at Steeple Hill earlier that day. Summer had been helping him, but at ten o'clock Tom had told her to go home. She'd gone to change, and then heard a crash from one of the autopsy suites. Running in, she'd found Tom semi-conscious on the floor.

'It was lucky she was still there,' Paul told me. 'If she hadn't been he could've been lying there for hours.'

He and Sam had been coming out of the Emergency Department as I arrived, blinking as they emerged into the bright sunlight. Sam looked calm and dignified, walking with the stately, leaned-back balance of late pregnancy. By comparison Paul seemed haggard and drawn with worry. He'd only found out about the heart attack when Mary had phoned him from the hospital that morning. Tom had undergone an emergency bypass during the night and was still unconscious in intensive care. The operation had gone as well as it could under the circumstances, but there was always the danger of another attack. The next few days were going to be critical.

'Do we know anything else yet?' I asked.

Paul raised a shoulder. 'Only that it was a massive attack. If he hadn't been so close to Emergency he mightn't have made it.'

Sam squeezed her husband's arm. 'But he did. They're doing every-thing they can for him. And at least the CAT scan was OK, so that's good news.'

'They did a CAT scan?' I asked, surprised. That wasn't a routine diagnostic for heart attacks.

'For a while the doctors thought he might have had a stroke,' Paul explained. 'He was confused when he was brought in. Seemed to think something had happened to Mary instead of him. He was pretty agitated.'

'C'mon, hon, he was barely conscious,' Sam insisted. 'And you know how Tom is with Mary. He was probably just worried that she'd be upset.'

Paul nodded, but I could see he was still concerned. So was I. The confusion could have been caused by Tom's brain not receiving enough oxygen or by a blood clot from his misfiring heart. A CAT scan should have shown up any obvious signs of a stroke, but it was another worrying factor, even so.

'Lord, I just wish I'd not been away yesterday,' Paul said, his face lined.

Sam rubbed his arm. 'It wouldn't have made any difference. You couldn't have done anything. These things happen.'

But this needn't have. I'd been berating myself ever since I'd heard the news. If I'd bitten my tongue instead of provoking Hicks, the pathologist might not have been so hell bent on having me thrown off the investigation. I could have taken some of the workload from Tom, might even have spotted the danger signs of the impending heart attack and done something about it.

But I hadn't. And now Tom was in intensive care.

'How's Mary?' I asked.

'Coping,' Sam said. 'She's been here all night. I offered to stay with her, but I think she'd rather be alone with him. And their son might be flying in later.'

'Might?'

'If he can tear himself away from New York,' Paul said bitterly.

'Paul . . .' Sam warned. She gave me a small smile. 'If you want to say hello I'm sure Mary would appreciate it.'

I'd known Tom would be too ill for visitors, but I'd wanted to come anyway. I started to go inside, but Paul stopped me. 'Can you stop by the morgue later? We need to talk.'

I said I would. It was only just starting to dawn on me that he was effectively the acting director of the Forensic Anthropology Center. The promotion didn't seem to give him any pleasure.

The clinical smell of antiseptic hit me as soon as I stepped inside the emergency department. My heart raced as it sparked a flashback to my own time in hospital, but I quickly quelled the memory. My footsteps squeaked on the resin floor as I made my way along the corridors to the intensive care unit where Tom had been taken. He was in a private room. There was a small window in the door, and through it I could see Mary sitting next to his bed. I tapped lightly on the window. At first she didn't seem to have heard, but then she looked up and beckoned me in.

She'd aged ten years since I'd been to their house for dinner two nights ago, but her smile was as warm as ever as she moved away from the bedside.

'David, you needn't have come.'

'I only just heard. How is he?'

We both spoke in a low whisper, even though there was little chance of disturbing Tom. Mary made a vague gesture towards the bed.

'The bypass went well. But he's very weak. And there's a danger he might have another attack . . .' She broke off, moisture glinting in her eyes. She did her best to rally. 'You know Tom, though. Tough as old boots.'

I smiled with a reassurance I didn't feel. 'Has he been conscious at all?'

'Not really. He came round a couple of hours ago, but not for long. He still seemed mixed up over who was in hospital. I had to reassure him that I was all right.' She smiled, tremulously, her anxiety showing through. 'He mentioned you, though.'

'Me?'

'He said your name, and you're the only David we know. I think he wanted me to tell you something, but I could only make out one word. It sounded like "Spanish".' She looked at me hopefully. 'Does that mean anything to you?'

Spanish? It seemed like more evidence of Tom's confusion. I tried to keep my dismay from my face. 'Nothing I can think of.'

'Perhaps I misheard,' Mary said, disappointed. She was already glancing towards the bed, obviously wanting to get back to her husband.

'I'd better go,' I said. 'If there's anything I can do . . .'

'I know. Thank you.' She paused, frowning. 'I almost forgot. You didn't call Tom last night, did you?'

'Not last night. I spoke to him yesterday afternoon, but that was about four o'clock. Why?'

She gestured, vaguely. 'Oh, it's probably nothing. Just that Summer said she heard his cell phone ring right before he had the attack. I wondered if it was you, but never mind. It can't have been anything important.' She gave me a quick hug. 'I'll tell him you stopped by. He'll be pleased.'

I retraced my steps and went back outside. After the oppressive quiet of the ICU the sun felt glorious. I tilted my face up to it, breathing in the fresh air to clear the smell of illness and antiseptic from my lungs. I felt ashamed to admit it even to myself, but I couldn't deny how good it felt to be in the open again.

Mary's words came back to me as I walked back to my car. What was it Tom had said? *Spanish.* I puzzled over it, wanting it to make some sort of sense rather than be further evidence of his confusion. But try as I might I couldn't think what it could mean, or why he should have wanted her to tell me.

174

Preoccupied with that, it was only when I was driving away that I remembered what else Mary had told me.

I wondered who might have been phoning Tom at that time of night.

The pan has boiled dry. You can see the tendrils of smoke coming from it and hear its contents hissing as they start to burn. But it's only when the smoke begins to cloud above the stove that you finally rouse yourself from the table. The chilli is blackened and hissing with heat. The stink must be intense, but you can't smell anything.

You wish you were as immune to everything.

You pick up the pan but let it drop again as the metal handle stings your hand. 'Sonofabitch!' Using an old towel, you lift it from the cooker and carry it to the sink. Steam hisses as you run cold water into it. You stare down at the mess, not caring one way or the other.

Nothing matters any more.

You're still wearing the uniform, but now it's sweat-stained and creased. Another waste of time. Another failure. And yet you'd come so close. That's what makes it so hard to stomach. You'd watched from the shadows, heart hammering as you'd made the call. You'd worried your nerves might give you away, but of course they hadn't. The trick is to shock them, to tip them off balance so they don't think clearly. And it had gone just as you'd planned. It had been almost pathetically easy.

But as the minutes ticked by he still didn't appear. And then the ambulance had arrived. You could only watch helplessly as the paramedics ran into the building and returned with the unmoving figure strapped to the trolley. Then they'd bundled it inside and driven him away.

Out of your reach.

It isn't fair. Just when you were on the point of triumph, of parading your superiority, it's been snatched away. All that planning, all that effort, and for what?

For Lieberman to cheat you.

'Fuck!'

The pan clatters against the wall as you fling it across the kitchen, leaving a trail of water and swinging flypapers. You stand with your fists balled, panting, desperate to feed the anger because behind it is only fear. Fear of failure, fear of what to do next. Fear of the future. Because, let's face it, what do you have to show for all the years of sacrifice? Worthless photographs. Images that show only how close you came, that have captured nothing but one near miss after another.

Tears sting your eyes at the injustice. Tonight should have gone some way towards countering the despair that's built up as one disappointment after another has emerged from the developing tray. Taking Lieberman would have made up for some of that. Would have shown that you're still better than the false prophets who claim to know it all. You deserve that much, at least, but now even that has been snatched away. Leaving you with what? Nothing.

Only the fear.

You close your eyes as you're blasted by an image from childhood. Even now you can still feel the shock of it. The chill from the big, echoing room soaking into you as you step through the doorway. And then the stink. You can still recall it, even though your sense of smell is long since defunct, an olfactory memory like the phantom tingling of an amputated limb. You stop, stunned by what you see. Rows of pale, lifeless bodies, drained of blood and life. You can feel the pressure of the old man's hand as he grips your neck, indifferent to your tears.

'You want to see somethin' dead, take a good look! Nothin' special about it, is there? Comes to us all, whether we want it or not. You as well. Take a good long look, 'cause this is what it all comes down to. We're all just dead meat in the end.'

The memory of that visit gave you nightmares for years. You'd catch sight of your hand, see the bones and tendons covered by a thin layer of skin, and you'd break out in a clammy sweat. You'd look at the people around you and see those rows of pale bodies again. Sometimes you'd see your reflection in the bathroom mirror and imagine yourself as one of them.

Dead meat.

You'd grown up haunted by that knowledge. Then, when you were seventeen, you'd stared into a dying woman's eyes as the life — the light — went out of them.

And you'd realized that you were more than meat after all.

It had been a revelation, but over the years it had become harder to sustain your belief. You'd set out to prove it, but each disappointment had only undermined it more. And after all the work and planning, after all the risks, tonight's failure was almost too much to take.

Wiping your eyes, you go to the kitchen table where the Leica is partially disassembled. You'd started to clean it, but even that pleasure has turned to ashes. You slump down on to the chair and consider the pieces. Lethargically, you pick up the lens and turn it in your hand.

The idea comes from nowhere.

A sense of excitement starts to grow as it takes shape. How could you have overlooked something so obvious? It was there, staring you in the face all along! You should never have let yourself forget that you have a higher purpose. You'd lost sight of what was really important, let yourself become distracted. Lieberman was a dead end, but a necessary one.

Because if not for that you mightn't have realized what a rare opportunity you've been given.

You feel strong and powerful again as you contemplate what has to be done. This is it, you can feel it. Everything you've worked for, all the disappointment you've endured, it was all for a reason. Fate had dropped a dying woman at your feet, and now fate's intervened again.

Whistling tunelessly to yourself, you start to strip off the uniform. You've been wearing it all night. There's no time to take it to the laundry, but you can sponge it down and press it.

You're going to need it looking its best.

14

The overweight receptionist was on duty at the morgue when I arrived. 'You heard 'bout Dr Lieberman?' he asked. The sing-song voice was cruelly mismatched to his huge frame. He looked disappointed when I said I had, tutting and shaking his head so that his chins quivered like jelly. 'It's a real shame. Hope he's OK.' I just nodded as I swiped my card and went inside.

I didn't bother to change into scrubs. I didn't know if I'd be staying or not.

Paul was in the autopsy suite where Tom had been working. He was poring over the contents of an open folder on the workbench, but glanced up when I entered.

'How was he?'

'About the same.'

He gestured at the papers in the folder. The bright fluorescent lighting showed up the dark shadows under his eyes, making his tiredness more evident. 'I was going through Tom's notes. I know some of the background, but it'd help if you could bring me up to speed.'

Paul listened silently as I told him how the body discovered at the cemetery seemed almost certain to be Willis Dexter's, and how the remains exhumed from Dexter's grave seemed likely to belong to a petty thief called Noah Harper. I described the pink teeth we'd found on both Harper's remains and those of Terry Loomis, the victim in the mountain cabin, and how they appeared to contradict the blood loss and wounds on the latter's body. When I told him that the hyoid bones of both victims were intact, and so far there were no signs of knife cuts to the bones themselves, he gave a tired grin.

'It's either or. Cause of death could be strangulation or stabbing, but not both. We'll just have to hope we find definitive evidence for one or the other.' He looked down at the folder for a moment, then seemed to rouse himself. 'So, are you OK to carry on?'

It had been what I'd been hoping to hear earlier, but circumstances robbed the moment of any satisfaction. 'Yes, but I don't want to cause any more friction. Wouldn't it be better if someone else took over?'

Paul closed the folder. 'I'm not asking you to be polite. With Tom in hospital the faculty's going to be pretty stretched. I'll do what I can here, but the next few days are going to be hectic. Frankly, we could use the help, and it seems stupid not to use you when you've been involved from the start.'

'What about Gardner?'

'It's not his decision. This is a morgue, not a crime scene. If he wants our help I've made it clear that he can either trust our judgement or find someone else. And he isn't about to do that, not now he's lost Tom so soon after Irving was snatched on his watch.'

I felt a touch of guilt at the reminder. What with Tom's heart attack, I'd almost forgotten about the profiler.

'And what about Hicks?' I asked.

Paul's expression hardened. 'Hicks can go to hell.'

It was obvious he was in no mood to make concessions. The pathologist and Gardner would find him very different to work with from Tom, I thought.

'OK,' I said. 'Shall I carry on reassembling the exhumed remains?'

'Leave them for now. Gardner wants to confirm whether or not the bones from the woods are Willis Dexter's. Summer's made a start on unpacking them, so that's our priority for the moment.'

I turned to go, but then remembered what I wanted to ask. 'Mary said Tom tried to tell her something earlier. She said it sounded like "Spanish". Does that mean anything to you?'

'Spanish?' Paul looked blank. 'Doesn't ring any bells.'

I went to get changed after that. Paul had to go to an emergency faculty meeting, but said he'd be back as soon as he could. Summer was already in the autopsy suite where the remains from Steeple Hill had been taken, unpacking the last of the evidence bags from their boxes.

Somehow I wasn't surprised to find Kyle helping her.

Engrossed in their conversation, neither of them heard me enter. 'Hi,' I said.

Summer gave a cry and spun round, almost dropping the bag she'd just picked up. 'Omigod!' she gasped, sagging with relief when she saw it was me.

'Sorry. I didn't mean to startle you.'

She managed a shaky smile. Her face looked tear-stained and blotchy under the bleached hair.

'That's OK. I didn't hear you. Kyle was just lending a hand.'

The morgue assistant looked embarrassed but pleased with himself.

'How's it going, Kyle?'

'Oh, pretty good.' He waggled his gloved hand, the one he'd spiked on the needle. 'Healed up nicely.'

If the needle had been infected it wouldn't matter whether the wound was healed or not. But he'd be well enough aware of that himself. If he wanted to put on a brave face then I'd no intention of spoiling it.

'Summer was telling me about Dr Lieberman,' he said. 'How is he?'

'He's stable.' It sounded better than saying there was no change.

Summer looked as though she might cry. 'I wish I could have done more.'

'You did great,' Kyle assured her, his round face earnest. 'I'm sure he's going to be OK.'

Summer gave him a tremulous smile. He returned it, then remembered I was still there.

'Well, uh, I suppose I ought to get on. See you later, Summer.'

Her smile grew more dimpled. 'Bye, Kyle.'

Well, well. Perhaps something good might come out of this after all.

After he'd gone Summer seemed listless, without her usual exuberance as we finished unpacking the remains.

'Kyle's right. It's lucky you were here last night,' I told her.

The overhead lights glinted on her piercings as she shook her head. 'I didn't do anything. I feel like I should have done something *more*. CPR, or something.'

'You got him to hospital in time. That's the main thing.'

'I hope so. He seemed fine, you know? A little tired, perhaps, but that's all. He joked about buying me pizza to make up for keeping me late.' The ghost of a smile flickered across her face. 'When it got to ten o'clock he told me to go home. He said he wanted to check something before he left himself.'

I felt my curiosity stir. 'Did he say what?'

'No, but I guessed it was something to do with the remains from the cabin. I went to change and was on my way out when I heard his cell phone ring. You know that corny old ringtone he has?'

Tom would have had a few choice words to say at hearing Dave Brubeck's 'Take Five' described as 'corny'. But I just nodded.

'I didn't take much notice, but then there was this sudden crash from the autopsy suite. I ran in and found him on the floor.' She gave a sniff and quickly wiped her eyes. 'I dialled 911 and then held his hand and talked to him until the paramedics arrived. Telling him he was going to be all right, you know? I'm not sure he could hear me, but that's what you're supposed to do, isn't it?'

'You did well,' I reassured her. 'Was he conscious?'

'Not really, but he wasn't completely out. He kept saying his wife's name, like he was worried about her. I thought perhaps he didn't want her to be upset when she found out, so I told him I'd call her. I thought it might be better coming from me than the hospital.'

'I'm sure Mary appreciated it,' I said, although I knew that sort of news was never welcome, no matter who it came from.

Summer gave another sniff and wiped her nose. A little of her bleached hair had come loose from its Alice band, making her look younger than she was.

'I put his glasses and cell phone in a cupboard above the work-bench in your autopsy suite. I hope that's OK; they were on the floor and I didn't know what else to do with them.'

I was about to say that I'd make sure Mary got them, but then her words registered. 'You mean they were on the floor in my autopsy suite?'

'That's right. Didn't I say? That's where Dr Lieberman collapsed.'

'What was he doing in there?' I'd assumed Tom had been in his own autopsy suite when he'd had the heart attack.

'I don't know. Is it important?' she asked, looking worried.

I assured her that it wasn't. Even so, I was puzzled. Tom had been reassembling Terry Loomis's skeleton. Why would he have broken off to check on the exhumed remains?

The question continued to nag me as we took the skull and other bones from the cemetery to be X-rayed, but it was another hour before I had a chance to do anything about it. Leaving Summer to make a start on cleaning the remains, I went to see where Tom had collapsed.

The suite looked exactly as I'd left it. Only the skull and larger bones were set out on the examination table; the rest were still wait-ing their turn in plastic boxes nearby. I stood there for a while, trying to tell if anything had been moved or changed. But if it had I couldn't see it.

I went over to the cupboard where Summer had left Tom's glasses and phone. The glasses looked both familiar and forlorn without their owner. Or perhaps I was just colouring them with my own emotions.

I slipped them into my top pocket and was about to do the same with the phone when something occurred to me. I paused, feeling its weight in my hand as I tried to decide if what I had in mind was too much of an invasion of privacy.

That all depends what you find.

The phone had been left on overnight, but it still had plenty of power. It didn't take long to find where incoming numbers were stored. The most recent had been logged at 22.03 the previous night, just as Summer had said.

The same time as Tom's heart attack.

I told myself that it could be a coincidence, that the two events might not be connected. Still, there was only one way to find out.

The number was from a landline with a local Knoxville code. I keyed it into my own phone. I had enough doubts about what I was doing as it was without using Tom's. Even then I still hesitated. *You might as well try it. You've come this far.*

I rang the number.

There was a pause, then the engaged tone sounded in my ear. With a sense of anticlimax I rang off and left it a minute before trying again. This time I was connected. My pulse quickened as I waited for someone to answer.

But no one did. The phone rang on and on, repeating itself with monotonous regularity. Finally accepting that no one was going to pick up, I broke the connection.

There were any number of reasons why the line should have been busy one minute and unanswered the next. The person at the other end might have gone out, or decided to ignore an unknown caller. It was useless speculating.

Still, as I left the autopsy suite, I knew I wasn't going to rest until I found out.

I was too busy for the rest of that day to think about trying the number again. The remains from Steeple Hill still had to be cleaned, but that was a relatively straightforward job. Scavengers and insects had already stripped any traces of soft tissue from them, so it was largely a matter of degreasing them in a detergent solution.

But we'd no sooner got them in the vats when the medical records of Noah Harper and Willis Dexter were delivered to the morgue. Knowing Gardner would want their IDs verified as soon as possible, I left Summer to finish cleaning and drying the bones while I turned my attention to that task.

Of the two, Dexter's identity proved the easier to confirm. The X-rays we'd taken that morning of the skull recovered from the woods showed identical fractures to those in X-rays taken at the mechanic's post mortem. It was what we'd expected, but now it was official: Willis Dexter wasn't the killer. He'd died in a car crash six months earlier.

That still left the question of whose body had been left in his grave.

There seemed little doubt that it was Noah Harper's, but we needed more than superficial similarities of age and race to be sure. Unfortunately, there were no post mortem or dental records to provide convenient identification. And while the eroded hip and ankle joints I'd found on the body from the casket would explain Harper's characteristic limp, there were no X-rays of them in his medical records. Medical insurance and dental care were obviously luxuries the petty thief couldn't afford.

In the end it was the childhood breaks in Harper's humerus and femur that identified him. They at least had been X-rayed, and although the grown man's skeleton was aged and worn, the long-healed fault lines in his bones remained constant.

By the time I'd satisfied myself as to the identities of both sets of remains, it was growing late. Summer had left a couple of hours earlier, and Paul had called to say that his meeting had overrun, so he wouldn't be able to make it back to the morgue after all. He'd got his priorities right, going home to his pregnant wife rather than working all hours. *Smart man.*

I would have liked to carry on working, but it had been a tiring day, emotionally as well as physically. Not only that, but I hadn't eaten since breakfast. Much as I might want to make up for lost time, starving myself was no way to go about it.

As I changed I called Mary to see how Tom was. But her phone was switched off, which I guessed meant she was still with him. When I called the ICU itself, a polite nurse told me he was stable, which I knew meant there was no change. I was about to put away my phone when I remembered the number I'd taken earlier from Tom's.

I'd forgotten all about it till then. I tried it again as I left the morgue, nodding goodnight to the elderly black man who now sat at reception.

The number was engaged.

Still, at least it showed that someone was home. I pushed open the heavy glass doors and stepped outside. Dusk was settling on the nearly empty hospital grounds, giving the evening a dying golden glow as I called the number once more. This time it rang. I slowed as I waited for someone to answer. *Come on, pick up.*

No one did. Frustrated, I ended the call. But as I lowered my mobile I heard what sounded like a distant after-echo.

A phone was ringing nearby.

It stopped before I could tell where it was coming from. I waited, but the only sounds were birdsong and the distant wash of traffic. Knowing I was probably over-reacting to what was in all likelihood just a coincidence, I called the number again.

A lonely ringing broke the evening's silence.

Perhaps thirty yards away, partially screened by a border of over-grown shrubs, was a public payphone. No one was using it. Still not quite believing this wasn't some fluke, I ended the call. The ringing stopped.

I redialled as I walked over. The payphone started ringing again. It grew louder as I approached, half a beat behind the tinnier version coming from my mobile. This time I waited until I was only a few feet away before I disconnected.

Silence fell.

The payphone was in a half-shell booth, open to the elements. Branches from the shrubs had grown round it, so that it seemed to be sinking into the greenery. I knew now why the line had either been busy or gone unanswered when I'd called. Hospitals were one of the few places where payphones were still in demand, visitors call-ing relatives or for taxis. Yet no one would bother to pick up if one rang.

I stepped into the booth without touching the phone. There was no doubt that someone had called Tom from here the night before, but I was at a loss as to why. Not until I looked back down the path I'd just walked along. Through the straggly branches of the shrubs I had a perfect view of the morgue entrance.

And of anyone who came out.

15

'So you think the killer called Dr Lieberman last night.'

Jacobsen's voice was completely inflectionless, making it impossible to know what she thought of the idea.

'I think it's possible, yes,' I said.

We were in the restaurant of my hotel, the half-eaten remains of my dinner congealing on the plate in front of me. I'd called Gardner from the hospital, finding his number in the address book of Tom's mobile. I'd anticipated his scepticism and readied my arguments for it. What I hadn't anticipated was that he wouldn't answer, and that I'd find myself having to explain to his voicemail service.

Rather than go into details, I'd said only that I thought the killer might have contacted Tom, and asked Gardner to call me. I'd assumed the TBI agent would want to see the payphone for himself, perhaps have it checked for fingerprints, although after it had been in use for another twenty-four hours I doubted there'd be much to find.

But there was no point waiting there until Gardner got my message and decided to call me. Feeling vaguely stupid, I'd gone to my car and driven back to my hotel.

It was almost an hour later before I heard anything. I'd just ordered dinner when my phone rang, but it was Jacobsen rather than Gardner on the other end. She asked for the number I'd taken from Tom's phone and told me to wait. The phone line went quiet, and I guessed she was passing the information on to Gardner. When she came back on she told me that she'd be with me in half an hour.

It was less than that when I looked up and saw her entering the restaurant. I pushed my plate away, my appetite suddenly gone. Jacobsen wore a black suit this time, the tailored skirt swishing with each stride as she made her way to my table. She could have been an ambitious young businesswoman except for the gun I glimpsed under her jacket as she sat down. There was no explanation of why Gardner hadn't returned my call or come himself, but I could guess.

Declining anything to eat or drink, she listened without comment as I explained in more detail about the call Tom had received.

Now I was starting to wish I hadn't bothered.

'Do you have Dr Lieberman's cell phone with you?' she asked.

I took it from my jacket and passed it over. I'd put it into my pocket at the last minute as I'd left my room. Just in case.

'Any news about Irving?' I asked, as Jacobsen examined the record of Tom's incoming calls.

'Not yet.' It was obvious that was all I was going to get. She copied the number into her own phone, then put it away without comment. 'What made you check Dr Lieberman's phone in the first place?'

'I was curious who'd called him. I wondered if it was connected with the heart attack.'

Her face was unreadable. 'You didn't think you might be prying?'

'Of course I did. But under the circumstances I didn't think Tom would mind.'

'Yet you didn't bother to ask anyone first?'

'Like who? Phone his wife while she's by his hospital bed?'

'I was thinking more of Dan Gardner.'

'Right. Because he values my opinion so much.'

Her smile seemed to take her by surprise as much as me. It lit up her entire face, changing her features from austerely attractive into ones that could grace the cover of a magazine. Then it was gone, leaving me wishing it had lasted longer.

'This is just conjecture,' she went on, the professional facade back in place. Though perhaps not so firmly as before. 'The call could have been made by anyone.'

'From a payphone right outside the morgue? At that time of night?'

She didn't answer. 'Have the doctors said when Dr Lieberman might be able to talk?'

'No. But probably not any time soon.'

We broke off as the waitress arrived to clear my plate and offer the dessert menu.

'Look, I'm going to have a coffee. Why don't you join me?' I said.

Jacobsen hesitated, glancing at her watch. For the first time a hint of tiredness leaked through.

'Maybe a quick one.' She ordered a latte, with skimmed milk and an extra shot of espresso.

'Are you sure you don't want anything else?' I offered.

'Coffee's fine, thank you,' she said, as though regretting even that much self-indulgence. I guessed that Jacobsen's blood sugar would always come off second best to self-discipline.

By tacit consent we put our discussion on hold while the waitress fetched our order. Jacobsen's fingers tapped restlessly on top of the banquette where we were sitting. Her nails were cut short, devoid of any polish.

'Are you from Knoxville originally?' I asked, to break the silence.

'A small town near Memphis. You wouldn't have heard of it.'

It was obvious I wasn't about to now, either. I tried again as the waitress set down the coffees.

'So what made you do a psychology degree?'

She hunched a shoulder. The movement seemed stiff and forced.

'It was an interest of mine. I wanted to pursue it.'

'But you joined the TBI instead? How come?'

'It was a good career move.'

She took a sip of coffee, closing the topic. *So much for getting to know her better.* I didn't think there was much point asking about a husband or a boyfriend.

'For the sake of argument, let's say you might be right about the phone call,' she said, lowering her cup. 'What would be the point? You're not suggesting someone deliberately caused Dr Lieberman's heart attack?'

'No, of course not.'

'Then why call him?'

Now we came to it. 'To lure him outside. I think Tom was going to be the next victim.'

The only outward sign of Jacobsen's surprise was her quick blink. 'Go on.'

'Tom seemed confused immediately after the heart attack, convinced that something had happened to Mary. Even at the hospital he had to be constantly reassured that she was all right. It was put down to the attack, but supposing it wasn't? Supposing someone called him and said his wife had had some kind of accident?'

The furrow was back between Jacobsen's eyes. 'So he'd rush out to go to her.'

'Exactly. When you get a phone call like that you forget about everything else. You don't worry about being careful or not going to your car alone. You drop everything and *go.*' I knew that only too well. The memory of hearing the policeman's voice telling me of my wife and daughter's accident still haunted me. 'At that time of night most of the hospital's pretty deserted, and the payphone where the call was made had a clear view of the morgue entrance. Anybody using it would have been able to see Tom come out.'

'Why not wait for him to finish work?'

'Because anyone planning to attack or abduct Tom wouldn't want to risk someone leaving with him. This way they'd be able to pick their moment, knowing he'd be alone and vulnerable.'

190

Jacobsen still wasn't convinced. 'They'd have to have got Dr Lieberman's cell phone number somehow.'

'Tom isn't shy about giving it out. Anyone could get it from his secretary at the university.'

'All right, but Dr Lieberman hasn't drawn attention to himself like Professor Irving. Why target him?'

'I've no idea,' I admitted. 'But you said yourself that whoever's behind this has a grandiose opinion of himself. Perhaps he felt that mechanics and petty thieves weren't getting him the attention he deserves.'

Jacobsen stared into space as she considered that. I made myself look away from the full lips.

'It's possible,' she conceded after a while. 'Maybe he's becoming more ambitious. Professor Irving could've whetted his appetite for more high-profile victims.'

'Unless Tom was the main target all along.'

I knew I was pushing my luck. Jacobsen frowned. 'There's no evidence to support that.'

'I know,' I agreed. 'It's just that I've been thinking about everything else the killer's done. Deliberately accelerated decomposition, pig's teeth substituted for human, and victims with apparently conflicting causes of death. All things guaranteed to get a forensic anthropologist scratching his head. Now it looks like Tom himself was nearly the next victim. Doesn't it strike you the killer could have had that in mind all along?'

She was still sceptical. 'Dr Lieberman isn't the only forensic anthropologist the TBI uses. There'd be no way anyone could be sure he'd be brought into this investigation.'

'Then perhaps the killer just wanted to set his cap against whoever was brought in, I don't know. But it's no secret that Tom's usually the TBI's first port of call. Or that he was planning to retire later this year.' *Sooner than that.* I pushed the thought of Tom and Mary's shattered plans away and pressed on. 'What if the killer saw this as his last window of opportunity to prove himself against one of the country's leading forensic experts? We know he arranged it so Terry Loomis's body would be found when the cabin rental expired, and Tom had only returned from a month's travelling the week before. That means the killer must have hired the cabin within a day or so of Tom's getting back. Supposing that wasn't just a coincidence?'

But I could see from Jacobsen's frown that I'd gone too far. 'Don't you think that's stretching things?'

I sighed. I wasn't sure myself any more. 'Perhaps. But then we're dealing with someone who planted hypodermic needles in a corpse six months before arranging to have it exhumed. Compared to that, making sure your next victim's going to be in town wouldn't be too difficult.'

Jacobsen was silent. I took a drink of coffee, letting her reach her own conclusions.

'It's reading an awful lot into one phone call,' she said at last.

'Yes,' I agreed.

'But I suppose it's worth looking into.'

Tension I'd not even been aware of till then bled out of me. I wasn't sure if I was relieved that a possible lead was being pursued, or just grateful to be taken seriously.

'So you'll check the payphone for fingerprints?'

'A crime scene team's there now, although after twenty-four hours I doubt they'll find anything.' Jacobsen's mouth quirked slightly at my surprise. 'You didn't think we'd just ignore something like that, did you?'

The *brrr* of her phone vibrating on the table saved me from having to answer. 'Excuse me,' she said, picking it up.

Feeling easier than I had all day, I drank my coffee while she went outside to take the call. I watched her through the glass doors, her features intent on whatever was being said. The conversation wasn't a long one. After less than a minute she came back inside. I expected her to make her excuses and leave, but instead she sat down at the table again.

She made no reference to the call, but there was a new coolness about her. The slight thaw I thought I'd detected earlier had vanished.

She moved the handle of her coffee cup minutely, repositioning it in its saucer. 'Dr Hunter . . .' she began.

'The name's David.'

She seemed caught off balance. 'Look, you ought to know . . .'

I waited, but she didn't go on. 'What?'

'It isn't important.' Whatever she'd been about to say, she'd thought better of it. Her eyes went to the almost empty beer glass that the waitress hadn't yet cleared. 'Forgive me for asking, but should you be drinking alcohol? Given your condition, I mean?'

'My condition?'

'Your injury.' She tilted her head quizzically. 'Surely you must have known we'd run a background check?'

I realized I was holding my coffee cup poised in mid-air. I carefully set it down. 'I hadn't given it much thought. And as for alcohol, I was stabbed. I'm not pregnant.'

The grey eyes regarded me. 'Does it make you feel uncomfortable talking about it?'

'There are pleasanter subjects.'

'Did you have any counselling after the attack?'

'No. And I don't want any now, thanks.'

An eyebrow cocked. 'I forgot. You don't trust psychologists.'

'I don't mistrust them. I just don't believe that talking about something is always the best way to deal with it, that's all.'

'Stiff upper lip, and all that?'

I just looked at her. A pulse of blood had started to tick away in my temples.

'Your attacker wasn't caught, was she?' she said, after a moment.

'No.'

'Does that worry you? That she might try again?'

'I try not to lose sleep over it.'

'But you do, though, don't you?'

I realized my hands were clenched under the table. They were clammy when I opened them. 'Is there a point to this?'

'I'm just curious.'

We stared at each other. But for some reason I felt calm now, as though I'd stepped over a threshold. 'Why are you trying to provoke me?'

Her gaze wavered. 'I'm only—'

'Did Gardner put you up to this?'

I don't know where the question came from, but when she looked away I knew I was right. It was only for a second, but it was enough.

'For Christ's sake, what is this? Are you *vetting* me?'

'Of course not,' she said, but without conviction. Now it was her turn to avoid my stare. 'Dan Gardner just wanted to assess your state of mind, that's all.'

'My state of *mind*?' I gave an incredulous laugh. 'I've been stabbed, I split up with my girlfriend, one of my oldest friends is lying in hospital, and everyone here seems convinced I'm incompetent. My state of mind's fine, thanks.'

Twin patches of colour burned on Jacobsen's cheeks. 'I apologize if I've offended you.'

'I'm not offended, just . . .' I didn't know what I was. 'Where is Gardner, anyway? Why isn't he here?'

'He's tied up with something else at the moment.'

I wasn't sure what annoyed me more, the fact he'd felt I needed assessing or that he hadn't deemed it important enough to do himself.

'Why bother with this now, anyway? The work's all but finished.'

The flush was fading from Jacobsen's cheeks. She stared pensively into her coffee, absently running a finger round the rim of the cup.

'A situation's developed at Steeple Hill,' she said.

I waited. The grey eyes met mine.

'York's disappeared.'

16

With lights burning in every window and TBI vehicles clustered outside, York's house had the starkly surreal look of a film set. It was in the grounds of Steeple Hill, hidden well away from the cemetery behind a fold in the pine woods. Like the funeral home itself, it was a low, rectangular block of concrete and glass, a failed attempt to transplant Californian 1950s modernism to the deep south. Once upon a time it might have been striking. Now, surrounded by the shadowy pinnacles of the pine trees, it just looked decayed and sad.

A crazed-paving path led to the front door, its slabs choked by straggly weeds. The crime scene tape that bracketed it gave the house an oddly festive air, although that impression was quickly dashed by the forensic agents searching it, ghost-like in their white overalls. At one side of the house, across an overgrown rectangle of lawn, a drive-way led to a garage. The door was raised, displaying a patch of oil-stained floor but no car.

That had disappeared along with its owner.

Jacobsen had briefed me on the drive over. 'We didn't see York as a realistic suspect for the homicide, otherwise we'd have arrested him sooner.' She'd sounded defensive, as though she were personally to blame. 'He fits the standard serial killer profile to some extent – right age, unmarried, a loner – and his inflated sense of self-importance is a typical narcissistic characteristic. But he doesn't have a criminal record, not even any warnings as a juvenile. No skeletons in his closet that we could find. Apart from the circumstantial evidence, there's nothing to link him to the actual killings.'

'The circumstantial evidence seemed pretty strong to me,' I said.

It was too dark in the car to see her blush, but I was sure she did. 'Only if you accept he deliberately incriminated himself by steering us towards the funeral home in the first place. That isn't unheard of, but his story about hiring a casual worker seemed to check out. We've found another former employee who claims to remember Dwight Chambers. It was starting to look as though Chambers might be a legitimate suspect after all.'

'So why arrest York?'

'Because holding him on public health charges would give us more time to question him.' Jacobsen looked uncomfortable. 'Also, it was felt that there were certain . . . advantages to taking a proactive approach.'

And any arrest looked better than no arrest. Politics and PR were the same the world over.

Except that York hadn't waited around to be arrested. When TBI agents went to pick him up that afternoon, there had been no sign of him either at the cemetery or his home. His car was missing, and when the TBI had forced entry into his house they'd found signs of hurried packing.

They'd also found human remains.

'We'd have discovered them sooner, except for a foul-up with the paperwork,' Jacobsen admitted. 'The original warrant only covered the funeral parlour and grounds, not York's private residence.'

'Are the remains recent?' I asked.

'We don't think so. But Dan would rather you see for yourself.'

That had shocked me even more than York's disappearance. It seemed that Paul had been unavailable. Sam was having a bad night.

They'd thought she was going into labour, and while that had proved to be a false alarm he wasn't prepared to leave her on her own.

So he'd told Gardner to ask me instead.

Paul had sounded tired and frazzled when I'd called him. Not that I doubted Jacobsen, but I wasn't about to go without speaking to him first.

'I've told Gardner I'll take a look first thing tomorrow, but if he wants an opinion tonight then he should ask you. Hope you don't mind,' he'd said. I told him I didn't, only that I was surprised Gardner had agreed. He gave a sour laugh. 'He didn't have much choice.'

He obviously hadn't forgiven Gardner for siding with Hicks against Tom. While Paul was too professional to let his personal feelings get in the way of an investigation, that didn't mean he couldn't turn the screw a little.

I wondered how Gardner felt about it.

Jacobsen hadn't stayed at Steeple Hill. After dropping me off she'd gone to check on the forensic team's progress with the payphone. I'd been directed to a van where I could change, and then made my way to the house.

Gardner was outside the front door, talking to a grey-haired woman in white overalls. He was wearing overshoes and gloves, and though he gave me a glance as I approached he didn't break off his conversation.

I stood at the bottom of the path and waited.

With a last terse instruction to the white-clad agent, Gardner finally turned to me. Neither of us spoke. His displeasure was almost palpable, but whatever he was thinking he kept to himself. He gave me a curt nod.

'It's upstairs.'

The house had the typical upside-down design of its style and era, so that the bedrooms were downstairs and the living quarters on the first floor. The once white walls and ceilings had been stained a dirty yellow by decades of cigarette smoke, and the same ochre patina clung to the doors and furniture like grease. Underlying the pervasive stink of stale tobacco was a musty smell of old carpets and unwashed sheets.

The sense of neglect and dilapidation was made worse by the turmoil of the search that was under way. Forensic agents were poring through drawers and cupboards, pulling out the detritus of York's life for examination. I felt their eyes on me as we went upstairs. There was an air of anticipation that I recognized from other crime scenes when a significant find had been made, but there was also open curiosity.

Word of my reinstatement had obviously got around.

Gardner led me up a staircase whose corners were felted with dust. The whole upper floor was open-plan, with kitchen, dining and living areas all combined. Most of the fittings looked original: partition shelf units and frosted glass cupboards straight from a 1950s advert for the domestic American dream.

But the furniture was a mishmash from the intervening decades. A rusted fridge hummed loudly in the kitchen, while an imitation chandelier with candle-shaped light bulbs hung over a scuffed dining table and chairs in the dinette. An overstuffed leather armchair sat in the centre of the living area, its split cushions patched with peeling electrical tape. Positioned in front of it was a huge flat screen TV, the only recent piece of furniture I'd seen.

There were more forensic agents busy up here. The house was in chaos, though it was hard to say how much was due to the search and what was the result of York's personal habits. Clothes were strewn about, and boxes of junk and old magazines had been pulled out of cupboards. But the sink and breakfast bar were invisible beneath dirty dishes, and crusted cartons of takeaway food lay where York must have dropped them.

Several of the search team broke off what they were doing to watch as Gardner led me across the room. I recognized the bulky form of Jerry on his hands and knees on the floor, poring through the drawers of a battered sideboard. He raised a gloved hand in greeting.

'Hi, doc.' The jowls of his face wobbled round his mask as he energetically chewed gum. 'Nice place, huh? And you should see his film collection. Porn paradise, all alphabetically listed. Guy really needs to get out more.'

Gardner had gone over to an alcove near the sink. 'So long as it's all still there when you're done.' There were chuckles, but I wasn't sure if he was joking. 'Through here.'

A walk-in cupboard was set in the alcove, its door wedged open. Its contents had been pulled out and lay spread around: boxes of chipped crockery, a plastic bucket with a split in its side, a broken vacuum cleaner. An agent knelt by a cardboard box of old photographic equipment: a worn SLR camera that had obviously seen better days, an old-fashioned flash unit and light meter, old photographic magazines, their pages faded and curling.

A yard or two away, isolated from the rest of the junk in a cleared space on the dusty linoleum, was a battered suitcase.

The lid was down but gaping, as though whatever was inside was too big for it to lie flat. Gardner looked down at it, making no attempt to approach too closely.

'We found it in the cupboard. Once we saw what was inside we left it alone until someone could take a look at it.'

The suitcase seemed too small to contain a human being. At least not an adult, but I knew that didn't mean anything. Years before I'd been called out to examine a grown man's body that had been crammed into a holdall even smaller than this. The limbs had been folded back on themselves, the bones broken and compacted into a shape no living contortionist could hope to achieve.

I squatted down beside it. The brown leather was scuffed and worn, but without the mould or staining I'd have expected if the remains had decomposed inside. That fitted with what Jacobsen had said about them not being recent.

'Can I take a look?' I asked Gardner.

'That's why you're here.'

Ignoring the acid in his voice I reached for the lid, conscious of everyone watching as I lifted it open.

The suitcase was full of bones. One glance was enough to confirm that they were human. There was what looked like an entire ribcage, against which a skull had been wedged, the mandible still connected so that it bore the hallmark grin. Looking at it, I wondered if Jacobsen's words in the restaurant had been intentional: *No skeletons in his closet that we could find.*

They'd found one now.

The bones were the same tobacco colour as the walls, although I didn't think cigarette smoke was responsible this time. They were clean, without any trace of soft tissue. I leaned closer and sniffed, but there was no real odour beyond the musty leather of the suitcase.

I picked up a rib that lay on the top. It was curved like a miniature bow. In one or two places I could see translucent flakes peeling away from the surface, like tiny fish scales.

'Any word yet on York?' I asked, as I examined it.

'We're still looking.'

'You think he left of his own accord?'

'If you mean was he abducted like Irving the answer's no. Irving didn't take his car or pack a suitcase before he disappeared,' Gardner said tersely. 'Now what can you tell me about these?'

I put the rib back down and took out the skull. The bones chimed together with almost musical notes as they shifted.

'They're female,' I told him, turning the skull in my hand. 'The bone structure's too delicate for a man. And she didn't die recently.'

'Tell me something I don't know.'

'OK,' I agreed. 'For a start she wasn't murdered.'

It was as though I'd suggested the earth was flat. 'What?'

'This isn't a murder victim,' I repeated. 'Look at how yellowed the bones are. This is *old*. Four or five decades at least. Perhaps more. You can see where it's been coated with some kind of stabilizer that's starting to flake off. I'm pretty certain it's shellac, which hasn't been used for years. And look at this . . .'

I showed him a small, neat hole drilled in the crown of the skull.

'That's where some sort of fixing used to be, so it could be hung up. Chances are this came from some lab or belonged to a medical student. Nowadays plastic models are used rather than actual skeletons, but you still come across real ones occasionally.'

'It's a *medical* skeleton?' Gardner glared down at it. 'What the hell is it doing here?'

I set the skull back in the suitcase. 'York said his father founded Steeple Hill back in the fifties. Perhaps it belonged to him. It's certainly old enough.'

'Goddammit.' He blew out his cheeks. 'I'd still like Paul Avery to take a look.'

'Whatever you like.'

I don't think Gardner even realized the implied slight. With a last disgusted look at the suitcase, he headed for the stairs. Closing the suitcase lid, I followed him.

'Bye, doc,' Jerry said, jaw still working. 'Another wasted trip, huh?'

As I passed the sideboard, I paused to look at the clutter of framed family photographs, a visual history of York's life. They were a mix of posed portraits and holiday snaps, the once bright summer colours washed out and faded. York was the subject of most: a grinning boy in shorts on a boat, an uncomfortable-looking teenager. An older, amiable-looking woman who looked like his mother was with him in most of them. Sometimes they were joined by a tall, tanned man with a businessman's smile who I took to be York's father. He wasn't in many, so I guessed he'd taken most of the photographs himself.

But the later shots were exclusively of York's mother, a progressively stooped and shrunken copy of her younger self. The most recent one showed her posing by a lake with a younger version of her son, frail and grey but still smiling.

There were no more after that.

I caught up with Gardner at the bottom of the stairs. So far he'd made no mention of the phone call Tom had received the night before. I wasn't sure if that was because he didn't think it was relevant, or if he just didn't want to acknowledge that I might have done something useful. But I wasn't going to leave without raising it.

'Did Jacobsen tell you about the phone booth?' I asked as we went along the hallway.

'She told me. We're looking into it.'

'What about Tom? If the call was meant to lure him outside he might still be in danger.'

'I appreciate you pointing that out,' he said, coldly sarcastic. 'I'll bear it in mind.'

I'd had enough. It was late and I was tired. I stopped in the hallway. 'Look, I don't know what your problem is, but you asked me to come out here. Would it kill you to at least be civil?'

Gardner turned and faced me, his face darkening. 'I asked you out here because I didn't have a hell of a lot of choice. Tom brought you into this investigation, not me. And excuse me if my manners aren't to your liking, but in case you haven't noticed I'm trying to catch a serial killer!'

'Well, it isn't me!' I flared back.

We glared at each other. We were by the front door, and through it I could see that the agents outside had stopped to stare. After a moment Gardner drew in a deep breath and looked down at the floor. He seemed to unclench himself with a visible effort.

'For your information, I arranged extra security for Tom straight away,' he said, in a tightly controlled voice. 'Purely as a precaution. Even if you're right about the phone call, I doubt that whoever made it is going to try anything while Tom's in a hospital bed. But I'm not about to take the chance.'

It wasn't exactly an apology, but I could live with that. The main thing was that Tom was safe.

'Thank you,' I said.

'You're welcome.' I couldn't decide if he was being facetious or not. 'Now, if that's all, Dr Hunter, I'll see you're taken back to your hotel.'

I started to go out, but I'd not even reached the front steps when someone called Gardner from inside the house.

'Sir? You should take a look at this.'

A forensic agent, overalls grubby with oil and dirt, had emerged from a door further down the hallway. Gardner glanced at me, and I knew what was going through his mind.

'Don't go just yet.'

He set off down the hallway and through the door. I hesitated, then went after him. I wasn't going to stand there like a schoolboy outside the headmaster's office until Gardner decided if he needed me or not.

The door was an internal entrance to the garage. The air smelled of oil and damp. A bare light bulb burned overhead, its weak glow supplemented by the harsher glare of floodlights. It was as cluttered in here as in the rest of the house, sagging cardboard boxes, mildewed camping gear and rusting garden equipment crowding round the bare area of concrete where York's car had stood.

Gardner and the crime scene agent were by an old steel filing cabinet. One of the drawers was pulled out.

'. . . at the bottom under old magazines,' the agent was saying. 'I thought at first they were just photographs, until I took a better look.'

Gardner was staring down at them. 'Jesus Christ.'

He sounded shocked. The other agent said something else, but I didn't pay attention. By then I could see what they'd found for myself.

It was a slim foolscap-sized box, the sort used for photographic paper. It was open, and the agent had fanned out the half-dozen or so photographs from inside. They were all black and white portraits, each a close-up of a man or woman's face from chin to forehead. They had been enlarged to almost full size, and the perfect focus had caught every feature, every pore and blemish, in sharp-edged detail; a split second preserved with unblurred clarity. Each face was contorted and dark, and at first glance their expressions were almost comical, as though each of the subjects had been caught on the point of a sneeze. But only until you saw their eyes.

Then you knew that there was nothing remotely comical about this at all.

We'd always suspected that there were more victims than the ones we knew about. This confirmed it. It hadn't been enough for York to torture them to death.

He'd photographed them dying as well.

Gardner seemed to notice I was there for the first time. He gave me a sharp look, but the rebuke I was half expecting never came. I think he was still too stunned himself.

'You can go now, Dr Hunter.'

A taciturn TBI agent drove me back to my hotel after I'd changed, but those contorted faces continued to haunt me as we drove through the dark streets. They were disturbing on a level that was hard to explain. Not just because of what they showed. I'd seen enough death in my time. I'd worked on cases before where murderers had taken trophies of their victims: a lock of hair or some scrap of clothing, twisted memento mori of the lives they'd claimed.

But this was different. York was no crazed killer, losing himself in the heat of some warped passion. He'd played us for fools all along, manipulating the investigation from the start. Even his exit had been timed perfectly. And the photographs weren't the usual trophies. They'd been taken with a degree of care and skill that spoke of a deliberate, clinical coldness. Of *control*.

That made them all the more frightening.

I didn't really need another shower when I got back to my room, but I had one anyway. The trip to York's house left me feeling unclean in a way that was more than skin deep. Symbolic or not, the hot water helped. So much so that I fell asleep almost the instant I turned out the light.

I was woken just before six by an insistent trilling. Still half asleep, I pawed for the alarm clock before I realized the noise was from my phone.

'Hello?' I mumbled, not properly awake.

The last vestiges of sleep fell away when I heard Paul's voice.

'It's bad news, David,' he said. 'Tom died last night.'

You cut it fine. You knew it wouldn't be long before the TBI agents arrived at the house, but you left it as long as you dared. Too soon and much of the impact would be lost. Too late and . . . Well, that would have spoiled everything.

It was a pity you didn't have more time. You hate feeling rushed, even though there was no avoiding it. You'd always known it would come to this. The funeral home had served its purpose. You'd planned it all out in advance; what you needed to take and what would be left behind. It had called for fine judgement and more than a little discipline. But that was OK.

Some sacrifices have to be made.

You're almost ready for the next stage now. All you've got to do is be patient. It won't be much longer. Just one final chore to nudge the last pieces into place, then the waiting will be over.

You admit to a few nerves, but that's a good thing. You can't let yourself be complacent. When the opportunity presents itself, you'll have to be ready to take it. You can't afford to waste chances like this. You know that better than anyone.

Life's too short.

17

In the end, all the precautions for Tom's safety had proved futile. Doctors and medical staff at the ICU had been warned of the need for extra vigilance, if not its reason, and a TBI agent had been stationed in the corridor outside his room. No one could have reached Tom without their knowing, and even if someone had, Mary had been at his side throughout.

None of which had prevented him going into cardiac arrest just after four o'clock that morning.

The medics had tried to resuscitate him, but his heart had resolutely refused to restart. *Stubborn to the end.* The thought circled aimlessly round my mind, refusing to settle.

I felt numb, still unable to take in what had happened. After I'd spoken to Paul I'd called Mary and mouthed the usual, useless words. Then I'd sat on my bed, at a loss as to what to do. I tried telling myself that at least Tom had died peacefully with his wife beside him, that he'd been spared whatever final ordeal had been inflicted on Irving. But it was scant consolation. York might not have physically killed him, but Tom was still a victim. Ill or not, he'd had a right to live the rest of his life in peace, however long it might have been.

He'd had that taken from him.

An image of York's face came to me, beaming with false servility as he'd enthusiastically pumped Tom's hand that morning at Steeple Hill. *Dr Lieberman, it's an honour, sir . . . I've heard a lot about your work. And your facility, of course. A credit to Tennessee.* He must have been laughing at us even then. Knowing what he had planned, hiding his greater guilt behind the petty misdemeanours evident at the cemetery.

I can't remember hating anyone as much as I hated York just then.

Moping in my hotel room wasn't going to bring Tom back, or help catch the man who'd killed him. I showered and dressed, then went to the morgue. It was still early when I arrived. My footfalls echoed as I walked down the empty corridor. The morgue's cold, tiled surfaces seemed even lonelier than usual. I would have welcomed the sight of a familiar face, but Paul had told me he had more meetings to get through first, and I doubted that Summer would be in any fit state to help out when she heard the news.

Kyle was there, at least. He was pushing a trolley along the corridor as I came out of the changing room, and greeted me with his usual enthusiasm.

'Hi, Dr Hunter. I've got to help with an autopsy this morning, but if you want any help after that, you just let me know.'

'Thanks, I will.'

He still loitered. 'Uh, will Summer be coming in later?'

'I don't know, Kyle.'

'Oh. OK.' He nodded, trying to hide his disappointment. 'How's Dr Lieberman?'

I'd guessed it was too soon for the news to have spread, but I'd been hoping he wouldn't ask. I didn't want to be the one to have to break it.

'He died last night.'

Kyle's face fell. 'He's dead? I'm sorry, I didn't know . . .'

'There's no reason why you should.'

I could see him searching for something to say. 'He was a nice man.'

207

'Yes, he was,' I agreed. There were worse epitaphs.

I tried to keep my mind blank as I went to the autopsy suite, wanting to focus on what I had to do. But it was impossible in an environment that I associated so much with Tom. When I passed the suite where he had been working, I paused, then went in.

It looked no different from the day before. Terry Loomis's skeleton still lay on the aluminium table, now almost fully reassembled. It was like any other autopsy suite, with no lingering trace of Tom's presence. I started to go back out, but then I saw the CD player still on the shelf next to the neat pile of jazz albums. That was when it really hit me.

Tom was dead.

I stood there for a while as the unalterable fact of it soaked in. Then, letting the weighted door swing shut, I went out and walked down the corridor to the autopsy suite where the bones of a petty thief were waiting.

The reassembly and examination of Noah Harper's skeleton should have been finished by now. The delay was no one's fault, but the task had been given to me and I felt responsible for how long it was taking. Now I was determined to complete it, if it meant staying all night.

Besides, I welcomed the distraction.

The cranium and larger bones of the arms and legs had been laid out on the table in an approximation of their anatomical position, but the rest had only been roughly sorted. I intended to reassemble the spinal column next, which was perhaps the most complex part of the process. The spine is essentially an articulated sheath that protects the cord of nerves at its centre. It's a perfect example of nature's ingenuity, a marvel of biological engineering.

But I was in no mood to appreciate it right then. Starting with the cervical vertebrae, I began carefully fitting the irregular knuckles of bone back together.

I didn't get far.

The cervical vertebrae that form the neck are smaller than the thoracic and lumbar vertebrae of the back. There are seven in all, numbered from the skull, each neatly dovetailing into those above and below. I fitted the first five together easily enough, but when I searched for the sixth I couldn't find it.

Come on, Hunter, concentrate. Exasperated, I went through the remaining vertebrae again. But the only cervical vertebra I could find was the wrong size and shape. It was clearly the seventh, not the sixth.

One was missing.

Which was impossible. Although it was badly decomposed, Noah Harper's body had been fully intact when we'd exhumed it. If one of his cervical vertebra had been absent we'd certainly have noticed.

So where was it?

With an odd sense of certainty, I went over to where the microscope stood on the workbench. I felt no surprise when I saw the small white object on the stage beneath the lens. If anything I should have realized before. I'd wondered what Tom had been doing in here when he'd had his heart attack.

Now I knew.

The image was blurred when I looked through the eyepiece. I adjusted the focus until the vertebra swam into view. It was as delicately fluted and spurred as coral, its porous surface appearing pitted under the magnification.

The hairline cracks looked as deep as a chasm.

Straightening, I took the piece of bone from under the microscope. The fractures were almost invisible to normal eyesight. There were two of them, one on each of the laminae, the slender bone bridges that link the main body of the vetebra to its more delicate neural arch.

Feeling strangely clear-headed, I set it down and went back down the corridor to the autopsy suite where Tom had been working. Going straight to Terry Loomis's skeleton, I picked up the sixth cervical vertebra from the examination table and held it up to the light. The fractures were even less obvious than on the laminae I'd just seen. But they were there all the same.

So that was it. I felt no satisfaction, only a sudden welling of sadness. This was Tom's discovery, not mine. I took out my phone and called Paul.

'I know how they were killed.'

'So it's definitely strangulation.'

Paul looked dispassionately down at the vertebra he was holding. We were in Tom's autopsy suite. I'd already shown him the fractures in Noah Harper's sixth cervical vertebra before bringing him in here to examine the matching cracks in Terry Loomis's.

'I can't see any other way you'd get such precise breaks,' I said. While a blow to the back of the neck could have broken the vertebra, the damage would have been much more extensive. And the chances of blunt trauma causing near identical injuries to two different victims was too remote to consider. No, these fractures were the result of something altogether more focused. More controlled.

That was a word that seemed to figure a lot with York.

'At least now we know for sure how Loomis and Harper came by the pink teeth,' Paul agreed. 'And it explains what Tom was doing in the other autopsy suite. He found the fractures in Loomis's vertebra and went to see if Harper's had them as well. That how you see it?'

'More or less.' And then York had phoned him while he was examining it under the microscope. I supposed there was an irony there, but wasn't sure what it was.

Paul gently set down the bone. 'Lord, it makes you want to weep.'

He sounded as exhausted as he looked. Tom's death had hit him hard, and the false alarm with Sam the previous night hadn't helped. He'd cut a faculty meeting short when I'd called him, and the strain of the last few days was evident as soon as he walked in. The lines around his eyes looked etched, and patches of blue-black bristles were already shading the pallor of his chin where he'd shaved in a rush.

He tried to stifle a yawn. 'Sorry.'

'Do you want to get some coffee?' I asked him.

'Later.' He made an effort to pull himself together. 'What about the cervical vertebrae from the remains in the woods? Have you checked them as well?'

'While I was waiting for you to get here. Two of them are missing, but the rest are all intact. Including the sixth.' That was no surprise. Willis Dexter had died in a car crash, not been murdered like Noah Harper or Terry Loomis.

'So we're looking at a steady pressure exerted on both victims' necks, powerful enough to fracture the laminae but without breaking the hyoid.' Paul held up his hands and considered them. 'Can you remember how big York's hands were?'

'Not big enough to do this.' The only thing I could recall about York's hands was their nicotine-stained fingers. But both Loomis and Harper had been grown men; it would have taken a huge span to wrap far enough round their necks to fracture the vertebrae. And that would more than likely have broken the hyoid as well.

'Most likely some kind of ligature or garrotte rather than manual strangulation,' Paul said. 'Whatever he used, it must've fastened around their necks at exactly the same point, causing identical damage to the same vertebra each time. Hard to say exactly what it was, though.'

'Tom had worked it out.'

He looked at me in surprise. 'He did?'

'Remember what he said to Mary when he was taken into hospital? *Spanish*. At the time we didn't know what he could have meant.'

It was a further sign of Paul's tiredness that it took him a moment to make the connection. 'Spanish *windlass*. Christ, I should have realized.'

211

So should I. Wrap a bandage or piece of cloth round a bleeding limb, then place a stick underneath and twist it. That was a Spanish windlass. At its most basic it was little more than an improvised tourniquet that could be wound tighter or loosened at will, a simple device that had saved countless lives.

But not the way York had used it.

I thought about the photographs the TBI had found in York's garage. The agonized expressions of his victims, their dark and swollen faces. Suffused with blood as York had incrementally tightened the windlass, steadily choking the life out of them.

And photographing it as it happened.

I pushed the images from my mind. 'York might not even have realized he was leaving any visible evidence at all. There'd be no way for him to know the laminae had been fractured. And even if he noticed the pink teeth, they're a pretty obscure phenomenon. He might not have realized their significance.'

'That still brings us back to the blood in the cabin,' Paul said. 'Loomis was strangled, so there's no way it's all his. So who the hell's is it?'

'Another of York's games, perhaps?' I said. The DNA analysis would tell us eventually, but I'd an idea that we might not have to wait as long as that.

Paul gave a weary shrug. 'I spoke to Gardner earlier. He didn't come right out and admit it, but they're obviously taking your theory about Tom seriously. The bottom line is they can't rule out that York might try for someone else on the investigation now he's screwed up his chances with him.'

I suppose that should have occurred to me, but somehow it hadn't. I'd been too wrapped up with what had happened to Tom to follow the idea to its logical conclusion.

'So what's Gardner going to do?'

'Not much he can do except warn people to be careful,' Paul said. 'He can't wrap everyone in cotton wool, and there isn't the manpower to do it even if he wanted to.'

'I'll consider myself warned.'

He smiled, but there wasn't much humour in it. 'Just keeps getting better and better, doesn't it? Turned out to be one helluva research trip for you.'

It had, but I was still glad I'd come. I wouldn't have missed the chance to work with Tom, regardless of how it had turned out.

'Are you worried?' I asked.

Paul's stubble rasped as he passed a hand across his face. 'Not really. York had surprise on his side before, but he's lost that now. I'm not saying I won't be careful, but I'm not going to spend my life looking over my shoulder in case some psycho decides to come after me.'

'You get used to it after a while,' I said.

He gave me a startled look, then broke out in a laugh. 'Yeah, I suppose you do at that.' He grew serious. 'Look, David, if you want to bow out, no one's going to blame you. This isn't your problem.'

I knew he meant well, but the reminder still felt like a slap. 'Perhaps not. But it sort of feels like it is.'

Paul nodded, then looked at his watch and grimaced. 'Sorry, but I've got to go. Another damn faculty meeting. Things should settle down in a day or two, but right now I need to be in two places at once.'

The silence of the autopsy suite seemed to press in around me after the door had closed behind him. I looked down at the partially assembled skeleton waiting on the examination table, and thought of Tom.

Clearing my mind, I went back to work.

I worked even later than I'd intended. Partly because I wanted to make up for lost time, but also because the thought of spending the evening alone back at my hotel held little appeal. As long as I was busy, I could hold off confronting the fact of Tom's death for a little longer.

But that wasn't the only thing that was bothering me. The feeling of oppression I'd felt after Paul's visit stubbornly refused to diminish. My senses seemed oddly heightened. The mortuary's chemical stink was underlaid with an indefinable biological odour, a faint hint of the butcher's slab. The white tiles and metal surfaces gleamed coldly in the harsh light. But it was the silence that I was most aware of. There was the distant hum of a generator, more felt than heard, the constant *plip* of a dripping tap. But, other than that, nothing. Normally I didn't even notice the quiet.

Now I felt it all around me.

Of course, I knew all too well what was the matter. Until Paul had mentioned it I'd never considered the possibility that York might target someone else from the investigation. My concern had been all for Tom, and even after what had happened to Irving, I'd blindly assumed that he was the only one under threat. But it was naive to think that York would stop with his death.

He'd just shift his priorities and carry on.

Paul hadn't really been involved with the investigation until now, but there were plenty of others who might satisfy York's apparent desire for high-profile victims. I wasn't arrogant enough to think that I was one of them. Even so, for the first time in days I found myself fingering my stomach, feeling the scar tissue under the cotton scrubs.

It was after ten before I finished. Noah Harper's bones revealed nothing else of significance, but then I hadn't expected them to. The fractured cervical vertebra had said enough. I changed and set off down the mortuary's main corridor. I seemed to have the place to myself. There was no sign of Kyle, but he would have finished his shift long ago. One of the fluorescent strips wasn't working, making the corridor dim. Up ahead I could see a thin beam of light seeping out from beneath the door of one of the offices. I was walking past when a voice came from inside.

'Who's there?'

I recognized the bad-tempered bark immediately. I knew the intelligent thing would be to walk straight past. Nothing I could say would change anything; it wouldn't bring Tom back. *Leave it. It isn't worth it.*

I opened the door and went in.

Hicks sat behind the desk, paused in the act of closing a drawer. It was the first time I'd seen him since the scene at the cemetery. Neither of us spoke for a moment. The lamp cast a low circle of light on to the desk, throwing the rest of the small office into shadow. The pathologist stared at me sullenly from the edge of it.

'Thought you were a diener,' he muttered. I saw the half-full tumbler of dark liquid in front of him and guessed I'd interrupted him putting away a bottle.

I'd gone in there intending to let Hicks know what I thought of him. But as I looked at him slumped behind the desk, my appetite for confrontation vanished. I turned to go.

'Wait.'

The pathologist's mouth worked, as though he were trying out unfamiliar words before he spoke them.

'I'm sorry. About Lieberman.' He studied the blotter on the desk, one fat index finger tracing an abstract pattern on it. I noticed that his cream suit looked rumpled and soiled, and realized he'd been wearing it every time I'd seen him. 'He was a good man. We didn't always get on, but he was a good man.'

I said nothing. If he was trying to appease a guilty conscience I wasn't going to help him.

But he didn't seem to expect me to. He picked up the tumbler and stared morosely into it.

'I've been doing this job for over thirty years, and you know what the worst of it is? Every time it happens to someone you know, it still surprises the hell out of you.'

He pursed his lips, as though puzzling over the fact. Then he raised the tumbler to his lips and emptied it. Reaching down with a small grunt he opened the drawer and produced a nearly full bottle of bourbon. For an awful moment I thought he was going to offer me a drink, propose some maudlin toast to Tom. But he only topped up his glass before putting the bottle back in the drawer.

I stood there, waiting to see what else he might say, but he stared into space as though he'd either forgotten I was there or wished I wasn't. Whatever urge had prompted him to talk seemed to have been exhausted.

I left him to it.

The encounter was unsettling. The comfortably black and white terms in which I'd seen Hicks had been undermined. I wondered how many other nights he'd sat alone in the small office, a lonely man whose life was empty except for his work.

It was an uncomfortable thought.

Tom's loss was a solid ache under my breastbone as I left the morgue and headed for my car. The night was cooler than usual, the damp chill a reminder that winter was still only recent history. My footsteps echoed off the darkened buildings. Hospitals were never truly abandoned, but when visiting hours had passed they could seem lonely places. And the morgue was always set well away from casual eyes.

It wasn't far to the car park, and I'd left my car in an open, well-lit area in its centre. But Gardner's warning whispered in my mind as I walked towards it. What had seemed safe in daylight now took on a wholly different aspect. Doorways were shadowy holes, the grassy spaces that I'd admired in the sunshine now fields of solid black.

I kept my steps regular and even, refusing to give in to the primal urge to hurry, but I was glad when I reached my car. I took out my keys and unlocked it while I was still a few paces away. I'd started to open the door before I realized there was something on the windscreen.

A leather glove had been slipped under one of the wipers, its fingers spread out on the glass. Someone must have found it on the ground and put it there for its owner to see, I thought as I went to remove it. A subliminal voice tried to warn me that it was the wrong time of year for gloves, but by then I'd already touched it.

It was cold and greasy, and far, far too thin for any leather.

I snatched my hand away and spun round. The darkened car park mocked me, silent and empty. Heart thumping, I turned back to the object on the windscreen. I didn't touch it again. It wasn't a glove, I knew that now. And it wasn't leather.

It was human skin.

18

Gardner watched as a crime scene agent lifted the windscreen wiper and carefully removed the scrap of skin with a pair of tweezers. He and Jacobsen had arrived twenty minutes ago, accompanied by the large van that was the TBI's mobile crime scene lab. Lights had been set up round the car, and the entire area taped off.

'You shouldn't have touched it,' Gardner said, not for the first time.

'If I'd realized what it was I wouldn't have.'

Some of my irritation must have leaked into my voice. Standing next to Gardner, Jacobsen took her eyes from the crime scene team dusting the car for fingerprints. She gave me a faintly worried look, the slight tuck visible between her eyebrows again, but said nothing.

Gardner, too, fell silent. He had a large manila envelope that he'd brought with him, although so far he'd made no mention of what it might contain. He watched, expressionlessly, as a forensic agent carefully placed the skin in an evidence bag. This was a different team from the one I'd seen before. I found myself wondering if they were on another job or just standing down for the night. Not that it mattered, but it was easier thinking about that than what this new development might mean.

Holding the bag carefully in a gloved hand, the agent brought it over. He raised it up so Gardner could get a better look.

'It's human, all right.'

I didn't need him to tell me that. The skin was dark brown in colour, with an almost translucent texture. It was obvious now that it was too irregular to be a glove, but the mistake was understandable. I'd seen this sort of thing often enough before.

Just not on the windscreen of my car.

'So does this mean that York's been skinning his victims?' Jacobsen asked. She was doing her best to appear unruffled, but even her composure had been shaken.

'I don't think so,' I said. 'May I?'

I held out my hand for the evidence bag. The forensic agent waited until Gardner gave a short nod before passing it across.

I held it up to the light. The skin was split and torn in several places, mainly across its back, but still retained a vague hand-like shape. It was soft and supple, and an oily residue from it smeared the inside of the plastic bag.

'It wasn't flayed off,' I told them. 'If it had been then it'd be in a flat sheet. This is split in places, but it's still more or less whole. I think it sloughed off the hand in one piece.'

There was no surprise on either Gardner's or the forensic agent's face, but I could see Jacobsen still didn't understand.

'Sloughed?'

'Skin slides off a dead body of its own accord after a few days. Especially extremities like the scalp and feet. And the hands.' I held up the evidence bag. 'I'm pretty certain that's what this is.'

She stared at the bag, her usual diffidence forgotten. 'You mean it slid off a corpse?'

'More or less.' I turned to the forensic agent, who'd been watching with a sour expression. 'Would you agree?'

He nodded. 'Good news is it's nice and soft. Saves us having to soak it before we lift the fingerprints.'

I felt Gardner looking at me, and knew he'd already made the connection. But Jacobsen seemed appalled.

'You can get *fingerprints* from that?'

'Sure,' the agent told her. 'Usually it's all dried and brittle, so you have to soften it up in water. Then you slip it on like a glove and take the prints like normal.' He held up his own hand and waggled it to illustrate.

'Don't let us keep you, Deke,' Gardner said. The agent lowered his hand, a little shamefaced, and went back to the car. Gardner tapped the manila envelope against his leg. The look he gave me was almost angry. 'Well? Are you going to say it or shall I?'

'Say what?' Jacobsen asked.

Gardner's mouth compressed into a thin line. 'Tell her.'

'We've been wondering how York managed to leave his victims' fingerprints at the crime scenes months after they were dead,' I said as she turned to me. I gestured at the car. 'Now we know.'

Jacobsen's frown cleared. 'You mean he's been using the skin from their hands? Wearing it like *gloves*?'

'I've never heard of it being done to plant fingerprints before, but that's how it looks. That's probably why Noah Harper's body was so badly decomposed. York wanted the skin from its hands before he switched it with Willis Dexter's.'

And then he'd waited a few more days before going back to the woods and collecting the sloughed skin from Dexter's hands as well. Scavengers wouldn't have bothered with scraps of drying tissue when they'd got the entire body to feed on. And if they had . . .

He'd just have used someone else's.

I felt a weary anger at myself for not realizing sooner. My subconscious had done its best to tell me, prompting the déjà vu at the sight of my wrinkled hands when I'd peeled off the latex gloves, but I'd ignored it. Tom had been right. He'd told me I should listen more to my instincts.

I should have listened to him as well.

Jacobsen took the evidence bag from me. Her expression was a mixture of disgust and fascination as she studied its contents.

'Deke said this wasn't dried out. Does that mean it must have come from a body recently?'

I guessed she was thinking about Irving. Although no one had actually said as much, we all knew that the profiler must be dead by now. But even if he'd been killed straight away, it would have taken longer than this for the skin to slough off. Whoever this had come from, it wasn't him.

'I doubt it,' I said. 'It looks like it's been deliberately oiled to preserve it and keep the skin supple . . .'

I stopped as something occurred to me. I looked over at the car windscreen, at the greasy smears left on the glass by the skin.

'Baby oil.'

Gardner and Jacobsen stared at me.

'The fingerprint on the film container in the cabin was left in baby oil,' I said. 'Irving thought it was proof that the killings were sexually motivated, but it wasn't. That's what York's been using to keep the sloughed skin supple. Its natural oils would have dried out, and he'd have wanted the fingerprints to be nice and clear. So he oiled it like old leather.'

I remembered Irving's mocking jibe. *Unless the killer has a penchant for moisturizing . . .* He'd been closer to the truth than he knew.

'If York's been harvesting his victim's fingerprints, how come he didn't take the skin from Terry Loomis's hands as well?' Jacobsen wanted to know. 'That was still in the cabin with the body.'

'If it hadn't been we'd have noticed and guessed what was going on,' Gardner said, self-reproach making his voice harsh. 'York wanted to pick his own time to let us know what he was doing.'

I watched the forensic agents carefully dust another part of the car with fingerprint powder. They were making a thorough job of it. For all the good it would do.

221

'So why now?' I asked.

Gardner looked across at Jacobsen. She shrugged. 'He's bragging again, telling us he isn't afraid of being caught. Obviously, he doesn't think our knowing this'll do us any good. Sooner or later we would've realized what he was doing anyway. This way he gets to stay in control.'

The other question remained unspoken. *Why me?* But I was afraid I already knew the answer to that.

Gardner looked down at the manila envelope he was holding. He seemed to reach a decision. 'Diane'll drive you to your hotel. Stay there till I call. Don't let anyone into your room; if someone says room service, make sure it is before you open the door.'

'What about my car?'

'We'll let you know when we're done with it.' He turned to Jacobsen. 'Diane, a word.'

The two of them walked out of earshot. Gardner did all the talking. I saw Jacobsen nodding as he handed her the envelope. I wondered what might be in it, but I couldn't raise much interest.

I looked back at the white-suited figures working on my car. The fine powder they were using to dust for fingerprints had dulled its paint, making it seem like something dead itself.

There was a bitter taste in my mouth as I watched them. I ran my thumb across the scar on my palm. *Admit it. You're scared.*

I'd been stalked by a killer once before. I'd come here hoping to put it behind me.

Now it was happening again.

It started to rain as Jacobsen drove me back to my hotel. Fat drops slid down the car windows in uneven bursts, swept away by the wipers only to reappear a moment later. Away from the hospital, the roads and bars were still busy. The bright lights and bustling streets were a relief, but I couldn't connect with their normality. I felt separated from them by more than the car window, aware that the reassurance they offered was illusory.

For once I was almost unaware of Jacobsen's closeness. It was only when she finally spoke that I dragged my thoughts back to the here and now.

'Dan says Loomis and Harper were strangled with some kind of ligature,' she said.

I stirred, surprised by the conversational gambit. 'Probably something called a Spanish windlass. A sort of tourniquet.' I explained how it worked.

'That'd fit in with what we know about York. He'd like the power something like that would give him. Literally life or death, and much more satisfying than killing someone straight away. It'd allow him to control the process, decide exactly when to exert enough pressure to kill his victim.' She gave me a quick glance. 'Sorry, that wasn't very tactful.'

I shrugged. 'It's all right. I've seen what York does. I'm not going to faint because he's playing mind games.'

'Is that what you think tonight was?'

'If he was serious about coming after me, why warn me in advance?' But even as I said it I realized I'd encountered another killer once who'd done exactly that.

Jacobsen wasn't convinced either. 'York needs to assert himself. To a narcissist like him, what happened with Dr Lieberman would've been a huge loss of face. His self-esteem's going to demand something even more spectacular to make up for it. Warning his next victim in advance might be it.'

'I still can't see why York would bother targeting me. Tom and Irving were both well known. Why go from high-profile targets to a stranger no one here's heard of? It doesn't make any sense.'

'It might to him.' She spoke flatly, her gaze on the road. 'He saw you working with Dr Lieberman, don't forget. And you're British, a

223

guest at the facility. York might feel that someone like you might make a bigger splash than someone local.'

That was something I hadn't considered. 'I suppose I should be flattered,' I said, trying to make a joke of it.

I wasn't rewarded with a smile. 'I don't think you should take it lightly.'

Believe me, I'm not. 'Can I ask something?' I said, wanting to change the subject. 'Have you heard anything from the lab about the blood samples from the cabin?'

There was a beat before she answered. 'A full DNA analysis takes weeks.'

That wasn't what I'd asked, but her evasion told me I was on the right track. 'No, but they should have found out by now if the blood was human or not.'

At any other time I might have enjoyed her surprise. 'How did you know that?'

'Call it an educated guess. So it was from an animal, then?'

The darkened profile gave a nod. 'We only got the results this afternoon, but even before then we knew there was something not right about it. Forensics weren't convinced by the spatter patterns in the cabin, although York made a good job of faking them. So the lab ran a preliminary test which suggested the blood was non-human. But we still had to wait until they'd extracted the DNA before we could be sure.'

'What was it? Pig's blood?'

I could see the white of her teeth in the darkness as she smiled. 'Now you're just showing off.'

Well, perhaps a little. 'It isn't as clever as it sounds,' I admitted. 'Once we'd confirmed that Terry Loomis had been strangled, then the blood obviously couldn't have been his. So the cuts on his body had to be post mortem, in which case most of the blood in the cabin had to have come from somewhere else.'

'I still don't see how you could know it was pig's blood . . .' she began, then answered herself. 'Oh, I get it. The teeth we found with Willis Dexter's body.'

'I'd wondered if the blood could be animal before then. But once I saw those I guessed it was probably from a pig as well,' I told her. 'Seems to be the sort of game York enjoys.'

Jacobsen fell silent. Her face was marbled by the rain running down the windows. In the slanting planes of yellow from the street-lights, her profile looked like a Grecian sculpture.

'I shouldn't really tell you this,' she said slowly. 'The blood samples from the cabin aren't the only results we've had. Noah Harper tested positive for Hepatitis C.'

God. Poor Kyle. Unlike the A and B strains, there was no vaccine for Hepatitis C. The virus wasn't necessarily fatal, but the treatment was time-consuming and unpleasant. And even then, there were no guarantees.

'Does Kyle know?' I asked, uncomfortably aware that it could easily have been me instead.

'Not yet. It'll be a while before he gets his own results from the hospital, and Dan didn't think there was any point worrying him.' She gave me a quick look. 'You understand this is strictly in confidence?'

'Of course.' For once I agreed with Gardner. There was still a chance Kyle might escape infection, but I wouldn't have wanted to stake my own life on so slim a bet.

We'd arrived at the hotel. Jacobsen found a parking space near the entrance. As she pulled in I saw her glance in the rear-view mirror, checking the cars behind us.

'I'll see you up to your room,' she said, reaching into the back seat for the manila envelope that Gardner had given her.

'There's no need.'

But she was already climbing out of the car. There was a new alert-ness about her as we went inside. Her eyes were constantly moving, flicking over the faces around us, checking for potential threats, and

I saw how she walked with her right hand held close to where her gun was concealed under her jacket. Part of me felt unable to take any of this seriously.

Then I remembered what had been left on my windscreen.

An elderly woman gave us a twinkling smile as she stepped out of the lift, and I could guess what she was thinking. *Just another young couple, on their way to bed after a day in the city.* It was so far removed from the truth it was almost funny.

Jacobsen and I stood side by side in the lift. We were the only passengers, and the tension between us seemed to increase with every floor. Our shoulders brushed lightly at one point, causing a quiet *snap* of static. She swayed away, just far enough to break the contact. When the doors opened she stepped out first, her hand slipping under her jacket to rest on the gun at her hip as she checked that the corridor was empty. My room was at the far end. I swiped my key card through the slot and opened the door.

'Thanks for escorting me.'

I was smiling as I said it, but she was all efficiency now. The barriers that had briefly come down in the car had gone back up.

'May I take a look in your room?'

I was going to tell her again there was no need, but I could see I'd be wasting my time. I stepped aside to let her in.

'Feel free.'

I stood by the bed while she searched. It wasn't a big room, so it didn't take her long to satisfy herself that York wasn't hiding in it. She was still carrying the manila envelope from Gardner, and when she'd finished she brought it over to where I waited. She stopped a few feet away, her face a perfect mask.

'One more thing. Dan wanted me to show you these.' She busied herself opening the envelope. 'There was a security camera over the road from the hospital payphone. We pulled the footage from the time the call was made to Dr Lieberman.'

She handed me a thin sheaf of photographs. They were stills from a CCTV camera: low quality and grainy, with the date and time printed at the bottom. I recognized the stretch of road where the phone booth was situated. One or two cars and the boxy white shape of an ambulance were partially visible in the foreground, blurred and out of focus.

But I was more concerned with the dark figure that was caught turning away from the payphone. The image quality was so poor it was impossible to make out its features. The head was bowed, the face no more than a white crescent that was all but hidden by a dark, peaked cap.

The other photographs showed more of the same, the figure hurrying across the road, shoulders hunched and head down. If anything it was even less clear in those.

'The lab's trying to clean up the images,' Jacobsen told me. 'We can't say for sure that it's York, but the height and build look about right.'

'You aren't just showing me these out of courtesy, are you?'

'No.' She looked at me unflinchingly. 'If you're York's next target Dan felt you ought to know what he might do to try to get near you. The dark clothes and cap could be some kind of uniform. And if you look on his hip there's something that looks like a flashlight. It's possible he tries to pass himself off as a police officer or some other authority figure who— Dr Hunter? What is it?'

I was staring at the photograph as the memory fell loose. *Flashlight . . .*

'A security guard,' I said.

'I'm sorry?'

I told her about being stopped in the car park a few nights earlier. 'It's probably nothing. He just wanted to know what I was doing there.'

Jacobsen was frowning. 'When was this?'

I had to think back. 'The night before Irving was abducted.'

'Did you get a good look at him?'

'He kept the torch pointed at my face. I couldn't see him at all.'

'What about anything else? His mannerisms or voice?'

I shook my head, still trying to recall. 'Not really. Except . . . well, his voice sounded . . . *odd*, somehow. Gruff.'

'Like he was disguising it?'

'It's possible.'

'And you didn't mention this to anyone?'

'I didn't think anything of it at the time. Look, it probably was just a security guard. If it was York why did he let me go?'

'You said yourself it was the night before Professor Irving disappeared. Maybe he had other plans.'

That silenced me. Jacobsen put the photographs back in the envelope.

'We'll check with hospital security, see if it was one of their people. In the meantime, keep your door locked when I've gone. Someone'll be in touch tomorrow morning.'

'So I've got to just wait here until I hear from you?'

She was all stone again now. 'It's in your own interests. Until we know how we're going to play this.'

I wondered what she meant by that, but let it go. Any decision would come from Gardner or above, not her. 'Do you want a drink before you go? I don't know how well stocked the minibar is, but I could order coffee or—'

'No.' Her vehemence seemed to surprise both of us. 'Thanks, but I need to get back to Dan,' she went on more calmly. But the flush spreading from the base of her throat told another story.

She was already heading for the door. With one last reminder for me to keep it locked, she was gone. *What was that about?* I wondered if she could have read too much into my offer of a drink, but I was too tired to worry about it for long.

I sank down on the edge of the bed. It seemed impossible that it was only that morning I'd heard of Tom's death. I'd intended to call Mary again, but it was too late now. I put my head in my hands.

Christ, what a mess. Sometimes it seemed I was dogged by ill luck and disaster. I wondered if events would have followed the same track if I'd never come out here. But I could almost hear what Tom would say to that: *Stop beating yourself up, David. This would have happened no matter what. You want to blame someone, blame York. He's the one responsible.*

But Tom was dead. And York was still out there.

I stood up and went to the window. My breath fogged the cool glass, reducing the world outside to indistinct yellow smudges in the darkness. When I wiped my fist across the pane, it reappeared with a squeak of skin on glass. The street below was a bright neon strip, car headlights creeping along in a silent ballet. All those lives, busily going about their own concerns, all indifferent to each other. Watching them made me acutely aware of how far from home I was, how much I didn't belong.

Whether you belong or not, you're here. Get on with it.

It occurred to me that I still hadn't eaten. Turning away from the window, I reached for the room service menu. I opened it but only glanced at the gushing descriptions of fast food before tossing it aside. All at once I couldn't stand to be in the room any longer. York or no York, I wasn't going to hide away until Gardner decided what to do with me. Snatching up my jacket, I took the lift back down to the lobby. I only intended to go to the hotel's late-night bar to see if they were still serving food, but I found myself walking straight past. I didn't know where I was going, only that I needed to be some-where else.

Outside, the rain had stopped, but the air was still freshened by its recent fall. The pavement was slick and shiny. My shoes raised small splashes as I set off down the street. The skin between my shoulder blades twitched, but I resisted the impulse to look behind me. *Come on then, York. You want me? Here I am!*

But my bravado soon burned itself out. When I came to a diner that was still open I went inside. The menu was mainly burgers and fried chicken, but I didn't care. I ordered at random and handed the menu back to the waitress.

'Anythin' to drink?'

'Just a beer, please. No, wait – Do you have any bourbon? Blanton's?'

'We got bourbon, but just Jim or Jack.'

I ordered a Jim Beam with ice. When it arrived I took a slow drink. The bourbon traced a gentle fire down my throat, easing away the lump that had formed there. *Here's to you, Tom. We'll get the bastard soon, I promise.*

For a while I almost believed it myself.

The straps and cogs gleam in the lamplight. You polish them after every time, waxing the leather until it's soft and supple and the tooled steel gleams. There's no real need. It's an affectation, you know that. But you enjoy the ritual. Sometimes you think you can almost smell the warm beeswax scent of the saddle polish; probably just a faint trace memory, but it soothes you all the same. And there's something about the sense of preparation, of ceremony, that appeals. Reminds you that what you're doing has a purpose; that the next time might be the one. And this time it will be.

You can feel it.

You tell yourself not to get your hopes up as you lovingly burnish the leather, but you can't deny the tingle of anticipation. You always feel it beforehand, when everything is possible and disappointment is still in the future. But this time it seems different. More portentous.

Special.

Leaving the skin on the car windscreen was a calculated gamble, but well worth it. They were bound to realize what you'd been doing eventually; better for it to be on your terms, when you can use it to good effect. You're still in control, that's the main thing. By the time they realize what's happening it'll be too late, and then . . .

And then . . .

But that's something you shy away from. You can't see that far ahead.

230

Better to stay focused on the job at hand, on the immediate objective.

It won't be long now.

You gently turn the winding mechanism, watching the leather strap tighten as the cogs turn smoothly, their teeth meshing with a clockwork whisper. Satisfied, you breathe on them before giving them a final rub. Your reflection stares back at you, distorted and unrecognizable. You stare at it, obscurely disturbed by thoughts that never quite break surface, then wipe it away with a sweep of the cloth.

Not much longer now, you tell yourself. Everything is in place and ready. The camera is loaded and in position, just waiting for its subject. The uniform is brushed and cleaned. Well, if not cleaned, exactly, at least clean enough to pass a first impression. And that's all you'll need.

It's all a matter of timing.

19

I was lingering over my second coffee in the hotel restaurant next morning when Gardner called.

'We need to talk.'

I glanced guiltily around the busy tables, conscious that he'd told me to stay in my room. I'd considered having my breakfast sent up, but in the bright daylight that didn't seem necessary. If York could spirit me out of the hotel in broad daylight then I was in real trouble anyway.

'I'm in the restaurant,' I said.

I felt Gardner's censure down the phone line. 'Stay there. I'm on my way over,' he told me, and hung up.

I sipped my cooling coffee, wondering if this was the last breakfast I'd be eating in Tennessee. I'd felt out of sorts all morning. I'd slept badly, waking with a heaviness I couldn't place at first. Then Tom's death came back to me, followed a moment later by the recollection of the skin left on my car.

It wasn't the best start to a day I'd ever had.

Gardner couldn't have been far away when he'd called, because he arrived within twenty minutes. Jacobsen was with him, looking as untouched and untouchable as usual. The late night seemed to have left no mark on her, but if her vitality held shades of Dorian Gray, then Gardner was the portrait in the attic. The senior agent looked worn out, the skin of his face a network of fine lines and grooves. I reminded myself that it wasn't just the pressure of the search for York that was weighing him down; Tom had been a friend of his as well.

But he held himself as straight as ever as he strode across to my table, Jacobsen a pace behind him.

'Can I get you a coffee?' I asked, as they sat down.

They both declined. Gardner glanced around the other tables to make sure no one could overhear.

'Security cameras show someone by your car at eight forty-five last night,' he said without preamble. 'It was too far away to pick out much detail, but the dark clothes and cap look the same as on the footage from the phone booth. Also, we checked with hospital security. It wasn't one of their employees you saw in the car park.'

'York.' There was a bitter taste in my mouth that had nothing to do with the coffee.

'We couldn't prove it in court, but we think so. We're still trying to identify the fingerprints we lifted from your hire car, but there're so many it isn't easy. And York probably wore gloves anyway.' Gardner shrugged. 'No luck with the sloughed skin, either. Its prints don't match either Willis Dexter's or Noah Harper's. From the small size it could be off a woman or an adolescent, but other than that we can't say.'

An adolescent. Christ. A skein of congealed milk lay on top of my coffee. I pushed it away from me. 'What about the photographs you found at York's house? Do you have any idea who the people in them are?'

Gardner looked down at his hands. 'We're checking them against the missing person database and unsolved homicides, but there's a lot to wade through. And it's going to be hard finding a match for them anyway.'

Remembering the contorted faces, I imagined it would. 'Have you any idea where York might be?'

'There've been a few unconfirmed sightings since we gave the press his details, but nothing definite. He's obviously got a hideout somewhere. He doesn't seem to have killed his victims either at his house or at Steeple Hill, so he must've taken them someplace else. Probably somewhere he can get rid of the bodies easily, or we'd have found others besides Loomis and Harper.'

With the Smoky Mountains on his doorstep, disposing of his victims' bodies wouldn't be difficult. 'According to Josh Talbot, for a swamp darner nymph to get caught up with Harper's body, it had to have been left near a pond or a slow-moving stream.'

'That narrows it down to almost the whole of East Tennessee.' Gardner gestured irritably. 'We've been checking out recorded sightings of swamp darners, but we need more to go on than that. Diane, why don't you tell Dr Hunter what you've come up with?'

Jacobsen tried to hide it, but there was a marked tension about her. I could see a pulse in the side of her throat, beating away in time to her excitement. I tore my eyes from it as she began to speak.

'I took another look at the photographs we found at York's house,' she began. 'They seem to have been taken when the victims were very close to death, perhaps at the actual point of death itself. I'd assumed they were just trophies York had collected. But if that's all they were, seeing how he'd strangled them you'd expect the victim's *throat* to be in the frame as well. It isn't, not in any of them. And if York just wanted to relive his kills, why not just record the whole thing on video? Why take such an extreme close-up of the victim's face, and in black and white at that?'

'Perhaps he's a photography buff,' I said.

'Exactly!' Jacobsen leaned forward. 'He thought he was being clever leaving Willis Dexter's fingerprint on the film canister, but he gave away more than he intended. Those photographs aren't just quick snapshots he's fired off. According to the lab they were taken in low light without a flash, using a very high speed film. To get a print of that quality under those conditions takes serious photographic know-how and equipment.'

'Wasn't there a thirty-five-millimetre camera at his house?' I asked, remembering the box of old photographic gear.

'The photographs weren't taken on that,' Gardner said. 'None of the equipment there had been used for years, so it was probably his father's. Judging from the pictures at the house York senior was an amateur photographer as well.'

I thought about the fading photographs on the sideboard. Something about them bothered me, but I couldn't think what.

'I still don't see why any of this is important,' I admitted.

'Because the photographs aren't just souvenirs to York. I think they might be central to what he's doing,' Jacobsen said. 'Everything we know about him suggests an obsession with death. His background, the way he treats his victims' bodies, his fixation with a forensic anthropologist like Dr Lieberman. Factor in these photographs of his victims *in extremis*, and it all points to one thing: York's a necrophiliac.'

Despite myself, I was shocked. 'I thought you said there was no sexual motivation?'

'There isn't. Most necrophiliacs are males with low self-esteem. They're drawn by the idea of an unresisting partner because they're terrified of rejection. That doesn't apply to York. If anything, he feels society doesn't appreciate him enough. And I doubt very much that he's attracted to his victims, dead or alive. No, I think his condition takes the form of *thanatophilia*. An unnatural fascination with death itself.'

This was getting into uncomfortable territory. I felt the first spike of a headache in my temples.

'If that's the case, why didn't he take the photographs when his victims were dead rather than as he killed them?'

'Because that wouldn't be enough. Over and above the

235

necrophilia, York's a pathological narcissist, remember. He's obsessed with *himself*. Most people are scared of dying, but to someone like him the notion of his own extinction must seem intolerable. He's been surrounded by death all his life. Now he's driven by a need to *understand* it.' Jacobsen sat back, her face solemn. 'I think that's why he kills, and why he takes photographs of his victims. His ego can't bear the thought that one day he's going to die himself. So he's looking for answers. In his own way he's trying to solve the mystery of life and death, if you like. And he's convinced himself that if he can take that definitive picture, catch the exact moment of death on film, it'll all become clear.'

'That's insane,' I protested.

'I don't think sanity is a prerequisite for serial killers,' Gardner commented.

He was right, but that wasn't what I meant. There was still no firm consensus on exactly when life ended. Stopped hearts could be resuscitated, and even brain death wasn't always conclusive. The idea that York thought he could capture the actual instant his victims died on film, let alone learn anything from it, disturbed me in ways I couldn't describe.

'Even if he managed it, what good does he think it'll do?' I asked. 'A photograph isn't going to tell him anything.'

Jacobsen gave a shrug. 'Doesn't matter. So long as York believes it then he'll carry on trying. He's on a quest, and it won't matter how many people he kills pursuing it. They're all just lab rats as far as he's concerned.'

The flush sprang up from her throat as she realized her mistake.

'I'm sorry, I didn't mean . . .'

'Forget it.' I might not like it, but I was no worse off for knowing what the situation was. 'From what you say, York's obviously been doing this for some time. Years, perhaps. God knows how many people he's killed already, without anyone knowing about it. He could have carried on like that indefinitely, so why the change?

What's made him suddenly decide to draw attention to what he's doing?'

Jacobsen spread her hands. 'Hard to say. But I'd guess it's precisely *because* he's been doing it for so long. You said yourself that what he's trying to do is impossible, and perhaps on some level he's started to realize that himself. So he's compensating, trying to make up for his failure by boosting his ego some other way. That's why he went after Dr Lieberman, a recognized expert in a field York probably regards as his own. In a way it's classic displacement – he's trying to avoid confronting his failure by reassuring himself that he is a genius after all.'

The headache had developed into a full-blown throb. I massaged my temple, wishing I'd brought some aspirin from my room.

'Why are you telling me this? Not that I don't appreciate it, but you haven't exactly been quick to share information before. So why the sudden change?'

Jacobsen glanced at Gardner. He'd seemed content to let her do most of the talking so far, but now he drew himself up almost imperceptibly.

'Under the circumstances it was felt that you'd a right to know.' He regarded me coolly, as though still assessing me even now. 'You've presented us with a problem, Dr Hunter. York was sending us a message by leaving the skin on your car. We can't ignore that. He's already abducted and in all likelihood murdered Alex Irving, and if not for the heart attack he'd probably have got Tom as well. I'm not about to let anyone else connected with the investigation be added to the tally.'

I looked down at my cold coffee, trying to keep my voice level. 'You can throw me off the investigation if you're like.' *Again.* 'But I'm not going back to the UK, if that's what you're thinking.'

It wasn't bravado. At the very least I wanted to stay for Tom's funeral. No matter what, I wasn't leaving without saying goodbye to my friend.

Gardner's chin jutted. 'That's not how it works. If we say you go, then you go. Even if it means having you escorted on to the plane.'

'Then that's what you'll have to do,' I retorted, my face growing hot.

The look he gave me said he'd like nothing better than to drag me to the airport himself. But then he let out a long breath.

'Frankly, it might be better for everyone if you were to go home,' he said sourly. 'But that wasn't what I had in mind. There could be certain . . . advantages if you stayed. At least then we'd know where to focus our attention.'

It took a moment for me to realize what he meant. I was too surprised to say anything.

'You'd be kept under constant surveillance,' Gardner went on, his manner businesslike now. 'You wouldn't be placed at any risk. We wouldn't ask you to do anything you were unhappy about.'

'And if I'm unhappy about the whole thing?'

'Then we'll thank you for your help and see you on to your plane.'

I felt an absurd urge to laugh. 'So that's my choice? I can stay, but only if I agree to be a stalking horse to draw out York?'

'That's your choice,' he said with finality. 'If you stay you'll need round-the-clock security. We can't justify that kind of expense when we could get you out of harm's way just by sending you home. Not without a good reason. But it's your decision. No one's twisting your arm.'

The brief relief I'd felt had dried up. Gardner was wrong; it was no decision at all. If I left then York would simply transfer his attentions to another victim.

I couldn't let that happen.

'What do I have to do?'

It was as though a bubble of tension had been pricked. A look of satisfaction flashed across Gardner's face, but Jacobsen was harder to read. For a second I thought I saw something like guilt cloud her eyes, but it had gone so quickly I could have been mistaken.

'For now, nothing. Just carry on as normal,' Gardner said. 'If York's watching I don't want him to realize anything's wrong. He'll expect us to take some precautions, so we won't disappoint him. We'll have a team parked outside the morgue and your hotel that he'll spot. But there'll be covert surveillance that he won't. You won't either.'

I nodded, as though all this was perfectly ordinary. 'What about my car?'

'We're done with it. Someone's bringing it to the hotel. They'll leave the keys at reception. We're still working on the details, but from tomorrow we'll have you drive out to places by yourself. You're going to be a tourist, taking walks by the riverside or on trails where you'll make an attractive target. We want to present York with an opportunity he can't ignore.'

'Won't he guess it's a trap if I start wandering off on my own?'

He gave me a flat look. 'You mean like you did last night?'

It took me a second to understand. I hadn't noticed anyone watching me when I'd left the hotel against his instructions, but I supposed I should have expected it. *So much for your grand gesture.*

'York might be suspicious at first, but we can be patient,' Gardner continued, having made his point. 'All he has to do is come out and sniff the air, and when he does we'll be there to take him.'

He made it sound easy. I'd been unconsciously rubbing my thumb across the scar on my palm. Realizing Jacobsen was watching me, I stopped and put my hands flat on the table.

'We're going to need you to work with us on this, Dr Hunter,' Gardner said. 'But if you'd rather you can be on a flight back home this afternoon. You can still change your mind.'

No, I can't. Conscious of Jacobsen's eyes on me I pushed back my chair and stood up.

'If that's all, I'd like to get to the morgue.'

I felt in a strange, unsettled mood all the rest of that day. There was too much to take in. Tom's death, finding myself next on York's list, and the prospect of being tethered out like a sacrificial goat the next day, all jostled for place in my mind. Each time I acclimatized myself to one I'd remember another, and be emotionally sandbagged all over again.

It was just as well that I didn't have anything demanding to do at the morgue. The more exacting tasks were finished, and all that remained was to sort and reassemble what little of Willis Dexter's skeleton had been recovered from the woods. That was purely routine, and wouldn't take long. Scavengers had made off with most of the bones, and the few that had been found were so badly gnawed that the hardest part was identifying what some of them were.

So there was nothing to distract my thoughts from following their vicious cycle. Nor was there anyone there I could talk to. Summer hadn't shown up that morning, although after Tom's death I hadn't really expected her to. There was little left for her to do anyway. But while I would have welcomed some company, I felt a coward's relief when one of the other morgue assistants told me that it was Kyle's day off. He'd still to learn about Noah Harper's positive Hepatitis C result, and just then I was glad I didn't have to face him.

Paul, too, was absent for most of the morning, tied up in the usual run of meetings. It was almost lunchtime before I saw him. He still looked tired, though not so much as the day before.

'How's Sam?' I asked, when he called into the autopsy suite.

'She's fine. No more false alarms, anyway. She's planning on seeing Mary this morning. Oh, and if you're not busy tonight you're invited for dinner.'

Under any other circumstances I would have been glad to accept. My social calendar wasn't exactly full, and the prospect of another night alone in my hotel was depressing. But if York was watching me the last thing I wanted was to involve Paul and Sam.

'Thanks, but tonight's not a good time.'

'Uh-huh.' He picked up a badly chewed thoracic vertebra and turned it in his fingers. 'I talked to Dan Gardner. He told me about

240

the skin left on your car last night. And that you'd volunteered to help catch York.'

I wouldn't have described it as *volunteered*, but I was glad Paul knew, all the same. I'd been wondering how much to tell him.

'It was either that or catch the next plane home.'

I was trying to make light of it. It didn't work. He set the vertebra back down on the examination table.

'You sure you know what you're getting into? You don't have to do this.'

Yes, I do. 'I'm sure it'll be fine. But you see why dinner isn't a good idea.'

'This is no time to be on your own. And I know Sam would like to see you.' He gave a grim smile. 'Trust me, if I thought there was any chance of putting her at risk I wouldn't be asking you. I'm not saying York isn't dangerous, but I can't see him being crazy enough to try anything now. Leaving the skin on your car was probably an empty threat. He had his big chance with Tom, and he blew it.'

'I hope you're right. But I still think we should leave it till some other time.'

He sighed. 'Well, it's your call.'

After he'd gone a wave of depression settled over me. I was almost tempted to phone and say I'd changed my mind. But only almost. Paul and Sam had enough going on in their lives as it was. The last thing I wanted was to take any trouble to their door.

I should have realized that Sam wasn't going to be put off that easily.

I was in the hospital's cafeteria, picking listlessly at a bland tuna salad and moodily contemplating the rest of the day, when she rang. She got straight to the point.

'So what's wrong with my cooking?'

I smiled. 'I'm sure your cooking's delicious.'

'Oh, it's the company, then?'

'It isn't the company either. I appreciate the invitation, really. But

I can't make tonight.' I hated being evasive, but I wasn't sure how much Sam knew. I needn't have worried.

'It's all right, David, Paul's told me what happened. But we'd still like to see you. It's thoughtful of you to be concerned, but you can't put yourself in quarantine until this creep's been caught.'

I gazed out of the window. People were walking past outside, absorbed in their own lives and problems. I wondered if York was out there somewhere. Watching.

'It's only for a few days,' I said.

'And if it was the other way round? Would you turn us away?'

I didn't know what to say to that.

'We're your friends, David,' Sam went on. 'This is an awful time, but you don't have to be alone, you know.'

I had to clear my throat before I could answer. 'Thanks. But I don't think it's a good idea. Not right now.'

'Then let's make a deal. Why don't you let this TBI guy decide? If he agrees with you, then you get to stay in your room and watch cable. If not, you come over tonight for dinner. OK?'

I hesitated. 'OK. I'll call him and see what he says.'

I could almost hear her smile down the phone line. 'I can save you the trouble. Paul checked with him already. He says he doesn't have any objection.'

She paused, giving me time to realize I'd been set up.

'Oh, and tell Paul to pick up some grape juice on your way over, will you? We're all out,' she added sweetly.

I was still grinning as I lowered the phone.

The traffic was bad heading out of Knoxville, but it eased the further from the city we went. I followed Paul, trying to keep his car in sight in the constantly streaming lanes. I switched on my radio, letting the anodyne music wash over me. But I still felt restless and on edge, glancing round every few minutes to see if I was being followed.

242

I'd called Gardner before we'd left. Not because I didn't trust Sam, but I still wanted to speak to him myself.

'Provided you take your own car and don't go walking off anywhere by yourself, I don't have a problem with it,' he'd said.

'So you don't think I'll be putting them at any risk?'

He sighed. I could hear the exasperation in his voice. 'Look, Dr Hunter, we want York to think you're behaving normally. That doesn't mean locking yourself in your hotel room every night.'

'But you'll have someone following me anyway?'

'Let us worry about that. Like I said, for now you just need to carry on as normal.'

Normal. There was precious little that was normal about the situation. Despite Gardner's reassurance, I'd left the mortuary through a back door rather than the main entrance. Then I'd driven round the hospital campus, meeting Paul at a different exit from the one I usually took. Even so, I couldn't shake the feeling that something was wrong. As I followed him away from the hospital, I repeatedly checked in the mirror. Nothing pulled out behind me. If the TBI or anyone else was there I couldn't see them.

Still, it was only when I'd merged with the homeward flow of evening traffic, becoming part of the metal river, that I began to accept that I wasn't being followed.

On the outskirts of Knoxville Paul stopped at a drive-by store for Sam's grape juice. He suggested I wait in my car, but I wasn't about to do that. So I went in with him, buying a bottle of Napa Valley Syrah I hoped would go with whatever Sam was cooking. The air was tainted with petrol and exhaust fumes as we walked back to the cars, but it was a beautiful evening. The sun was starting to set, throwing golden arms across the skyline, while the thickly wooded slopes of the Smoky Mountains purpled into dusk.

I gave a start as Paul swore and slapped at the back of his neck.

'Damn bugs,' he muttered.

He and Sam lived in a new lakeside development between

Knoxville and Rockford to the south. Part of it was still being built, piles of earth and timber giving way to manicured lawns and newly planted flower beds the further in we went. Their house was on a meandering side road that skirted the lake and curved round each property, giving a pleasing impression of space and privacy. The development still had a raw, unfinished look, but it had been well planned with plenty of trees, grass and water. It would be a good place to raise a family.

Paul turned into the driveway and pulled up behind Sam's battered Toyota. I parked on the road and climbed out to join him.

'We're still decorating the nursery, so don't mind the mess,' he said as we headed up the path.

I wouldn't. For the first time I felt glad I'd come, my spirits lighter than they'd been in days. Their house was set slightly back from the rest so that it had a larger garden. In a rare display of conservation and common sense, the builders had worked round a beautiful mature maple, turfing around it so the tree became a centrepiece. I remember thinking as we walked past that it would be ideal for a child's swing.

It's odd how these things stay with you.

'Paul? Wait up a second!'

The shout came from the neighbouring house. A woman was bustling across the lawn towards us. Tanned and trim, with too-bright blond hair coiffed into an elaborate bun, at first glance I'd have put her in her late fifties. But as she drew closer I revised that upwards, first to sixties and then seventies, as though she was ageing with every step.

'Oh, great,' Paul muttered under his breath. He mustered a dutiful smile. 'Hi, Candy.'

The name was too cute and too young, yet somehow suited her. She went to stand close to him, her poise that of an ageing model who doesn't realize her decade is over.

'I'm *so* glad I saw you.' Her too-white dentures gave her words a slight sibilance. She rested a liver-spotted hand on his arm, the veined skin as brown as old moccasins. 'I wasn't expecting to see you back so soon. How's Sam?'

'She's fine, thanks. Just a false alarm.' Paul started to introduce me. 'Candy, this is—'

'A false alarm?' Her face fell with dismay. 'Oh, Lord, not again. When I saw the ambulance, I felt *sure* it was for real this time!'

There was an instant when the evening seemed suspended. I could smell the freshness of the new grass and blossom, feel the first chill of night behind the spring warmth. The smooth weight of the wine bottle in my hand still held the promise of normality.

Then the moment shattered.

'What ambulance?' Paul looked more confused than concerned.

'Why, the one that came earlier. About four thirty, I guess.' The woman's painted smile was collapsing. Her hand fluttered to her throat. 'Surely someone *told* you? I thought . . .'

But Paul was already running towards the house. 'Sam? *Sam!*'

I quickly turned to the neighbour. 'Which hospital did she go to?'

She looked from where Paul had disappeared into the house to me, her mouth working. 'I . . . I didn't ask. The paramedic brought her out in a wheelchair, with one of those oxygen things on her face. I didn't want to get in the way.'

I left her on the path and went after Paul. The house had the smell of fresh paint and plaster, of new carpets and furniture. I found him standing in the middle of the kitchen, surrounded by gleaming new appliances.

'She's not here.' He looked stunned. 'Jesus Christ, why didn't somebody *call* me?'

'Have you checked your phone for messages?'

I waited as he did. His hand shook as he pressed the keys. He listened, then shook his head. 'Nothing.'

'Try the hospital. Do you know which one she'll have been taken to?'

'She's been going to UT Medical Center, but . . .'

'Call them.'

He stared at his phone, blinking like a man trying to wake up. 'I don't have the number. Christ, I should *know* it!'

'Try the operator.'

He was starting to come round now, his mind recovering from the initial shock. I stood by as he dialled the hospital, pacing during the agonizing transferrals. As he spelled out Sam's name for the third or fourth time, I could feel the presentiment that had dogged me all day steadily growing closer, until its presence filled the room.

Paul rang off. 'They don't know anything.' His voice was controlled, but the panic was close to the surface. 'I tried the Emergency Department as well. There's no record of her being admitted.'

Abruptly, he began jabbing at the keypad again. 'Paul . . .' I said.

'There must be some mix-up,' he mumbled, as though he hadn't heard. 'She must've been taken to another hospital . . .'

'*Paul.*'

He stopped. His eyes met mine, and I could see the fear in them, see the knowledge he was desperately trying to deny. But neither of us had that luxury any more.

I wasn't York's target. I never had been.

I was just the decoy.

20

The night that followed was one of the longest of my life. I called Gardner while Paul phoned the rest of the hospitals in the area. He must have known Sam wouldn't be at any of them, but the alternative was too terrible to accept. As long as the possibility remained, no matter how faint, he could cling to the hope that this was all just some mistake, that his world could still return to normal.

But that wasn't going to happen.

It took Gardner less than forty-five minutes to arrive. By then two TBI agents were there already. They'd appeared at the house within minutes, both in grubby work clothes as though they'd come from a building site. From the speed with which they arrived I guessed they must have been very close by, no doubt part of the covert surveillance that had been promised. Not that it had done any good.

Gardner and Jacobsen came into the house without knocking. Her features were carefully controlled; his were clenched and grim. He spoke briefly to one of the agents, a subdued murmur of voices, then turned to Paul.

'Tell me what happened.'

There was a tremor in Paul's voice as he went over it once more.

'Any sign of a disturbance? A struggle?' Gardner asked.

Paul just shook his head.

Gardner's eyes went to the coffee cups on the table. 'Have either of you touched anything?'

'I made coffee,' I said.

I saw in his face the accusation that I shouldn't have touched anything at all, but he didn't get the chance to voice it.

'To hell with the damn coffee, what are you going to *do*?' Paul burst out. 'This bastard's got my *wife*, and we're just sitting here talking!'

'We're doing everything we can,' Gardner said, with surprising patience. 'We've notified every police and sheriff's department in East Tennessee to look out for the ambulance.'

'*Notified* them? What about road blocks, for Christ's sake?'

'We can't flag down every ambulance on the off chance it might be York. And road blocks won't do any good when he's got several hours' head start. He could be over the state line into North Carolina by now.'

The anger drained from Paul. He slumped in his chair, his face ashen.

'This might be nothing. But I've been thinking about the ambulance,' I said, choosing my words carefully. 'Wasn't there one in the security camera footage? By the phone booth where York called Tom?'

It had been little more than a white shape in the foreground. I wouldn't have thought anything about it ordinarily, and I wasn't sure it was important even now. But I'd rather speak out of turn than stay silent and regret it.

Gardner obviously thought otherwise. 'It was a hospital, they have ambulances.'

'Outside emergency, perhaps, but not the morgue. Not at the main entrance, anyway. Bodies aren't taken in that way.'

He was quiet for a moment, then turned to Jacobsen. 'Tell Megson to look into it. And have the stills sent over.' He turned back to Paul as she hurried out. 'OK, I need to talk to the neighbour.'

'I'll come with you.' Paul got to his feet.

'There's no need.'

'I want to.'

I could see Gardner was reluctant, but he gave a nod. He went up in my estimation for that.

I was left alone in the house. The knowledge of how badly we'd been played for fools burned like acid. My noble gesture to Gardner, agreeing to offer myself as bait, now seemed nothing more than hubris. *God, have you got such a high opinion of yourself?* I should have realized that York wouldn't have bothered with me when there were far more tempting targets for the taking.

Like Sam.

The kitchen was in near-darkness, the daylight almost completely gone. I turned on the light. The new appliances and freshly painted walls seemed mocking in their optimism. I'd been in Paul's position once myself, but with one crucial difference. When Jenny had been abducted we'd known that her captor kept his victims alive for up to three days. But there was nothing to suggest that York kept his victims alive any longer than he had to.

Sam might be already dead.

Restless, I left the kitchen. A forensic unit was on its way to the house, but no one seriously expected them to find anything significant. Even so, I was still careful not to touch anything as I went into the lounge. It was a comfortable, cheery room: soft sofa and chairs, coffee table half covered with magazines. It was imprinted with Sam's personality far more than Paul's; thoughtfully designed, but still a room for living in rather than admiring.

I turned to go, and my eye fell on a small photo frame on the smoked glass cabinet. The picture was an almost abstract pattern of black and whites, but the sight of it was like a punch in the stomach.

It was a prenatal scan of Sam's baby.

I went back out into the hall. I stopped by the front door, visualizing what must have happened. *A knock on the door. Sam opening it, seeing a paramedic there. She'd be confused, convinced there was some mix-up. Probably smiled as she tried to explain the mistake. And then . . .* what? There were bushes screening the front door, the big maple tree in the garden further blocking it from view. But York wouldn't have taken any chances on being seen. So he'd have tricked or forced his way inside somehow, before quickly overpowering her and bundling her into the wheelchair.

Then he'd brazenly pushed her down the path to his waiting ambulance.

I noticed something on the floor by the skirting board, specks of white on the beige carpet. I bent down for a closer look, and jumped as the front door suddenly opened.

Jacobsen paused when she saw me crouching in the hall. I got to my feet and gestured at the white flecks.

'Looks like York was in a hurry. And no, I haven't touched anything.'

She examined the carpet, then the skirting board next to it. There were scuff marks on the woodwork.

'Paint. He must have caught the skirting board with the wheelchair,' she said. 'We'd wondered how York got Professor Irving out of the woods. It was a good half-mile to the nearest parking place. That's a long way to move a grown man, especially if he's unconscious.'

'You think he used a wheelchair then as well?'

'It'd explain a lot.' She shook her head, annoyed at the oversight. 'We found what looked like cycle tracks on the trails near where Irving went missing. It's a popular area with mountain bikers, so it didn't seem relevant at the time. But wheelchairs have similar tyres.'

And even if York had encountered anyone as he was pushing an unconscious Irving back along the trail, who would have thought anything of it? He'd just have looked like a carer taking an invalid out in the fresh air.

We went back into the kitchen. I saw Jacobsen looking at the half-full coffee percolator. Without asking I poured her a cup and topped up my own.

'So what do you think?' I asked, quietly, as I handed it to her.

'It's too soon to say . . .' she began, then stopped. 'You want me to be honest?'

No. I gave a nod.

'I think we've been two steps behind York all along. He fooled us into thinking you were his target, and walked in here while we were looking the other way. Now Samantha Avery's paid for our mistakes.'

'You think there's any chance of finding her in time?'

She looked into her coffee as though she could divine the answers there. 'York won't want to take long over this. He knows we're looking for him, and he'll be excited and eager. If he hasn't killed her already, she'll be dead before the night's out.'

I put my cup down, feeling suddenly nauseous. 'Why Sam?' I asked, although I could guess.

'York needed to reassert his ego after his failure with Dr Lieberman. We were right about that much, at least.' Jacobsen sounded bitter. 'Samantha Avery would've ticked all the boxes: the wife of Dr Lieberman's probable successor, and nearly nine months pregnant. That'd make her doubly attractive. It guarantees headlines and, if we're right about the photographs, it also feeds into York's psychosis. He's fixated on capturing the moment of death on film, believing that'll somehow reveal the answers he's looking for. So from his point of view, who could be a better victim than a pregnant woman, someone who's literally full of life?'

Christ. It was insane, and yet the worst of it was there was a twisted logic behind it. Futile and obscene, but there all the same.

'And what then? He isn't going to find the answers he's looking for by killing Sam.'

Jacobsen's face held a bleakness I'd not seen before. 'Then he'll tell himself she wasn't the right one after all and carry on. He'll know time's against him, no matter how much his pride says otherwise, and that's going to make him desperate. Maybe next time he'll go after another pregnant woman, or even a child. Either way, he won't stop.'

I thought of the tortured faces in the photographs and had a sudden image of Sam going through the same ordeal. I rubbed my eyes, trying to banish it.

'So what happens now?'

Jacobsen stared out of the window at the advancing night. 'We hope we find them before morning.'

Before the next hour was out, the evening's quiet had been rent apart. TBI agents descended on the sedate neighbourhood, knocking on every door in the hope of finding more witnesses. Plenty of people could recall seeing an ambulance that afternoon, but no one had noticed anything remarkable about it. Ambulances were self-explanatory. The sight of one might arouse morbid curiosity, but few people would question why it was there.

Certainly none of Sam and Paul's neighbours.

Gardner hadn't managed to learn anything more from Candy. All she could say for sure was that it had been a man of indeterminate age wearing a paramedic's uniform. Well, it *looked* like a uniform, she thought: dark trousers and a shirt with badges on it. And some sort of hat or cap that hid most of his face. A big man, she'd added, more hesitantly. White. Or perhaps Hispanic. Not black, at any rate. At least, she didn't think so . . .

It hadn't even struck her as odd that the ambulance driver had been alone. And she'd been able to offer even less information about the ambulance itself. No, of *course* she hadn't taken the licence number. Why should she? It was an *ambulance*.

'There were no obvious restraints, so Samantha must have been stunned or unconscious,' Gardner said, while Paul was on the phone to Sam's mother. 'It's possible he used some sort of gas, but I think the oxygen mask was probably just a prop to dissuade any watching

252

neighbours from intervening. Gas is too hit and miss, especially if someone's struggling, and York would've wanted to put her out as soon as possible.'

'He wouldn't use brute force,' Jacobsen said. 'If you knock someone unconscious there's a danger of concussion or brain damage, and York wouldn't want that. He needs his victims fully aware when he kills them. He wouldn't risk clubbing them over the head.'

'He did Irving's dog,' Gardner reminded her.

'The dog was incidental. He was after its owner.'

Gardner squeezed the bridge of his nose. He looked tired. 'Whatever. The fact is he obviously knocked Samantha Avery out somehow. But at least if he has to wait till she comes round, that might give us more time.'

I hated dispelling even that slight hope. 'Not necessarily. He only needs his victims unconscious long enough to get them into the ambulance. After that it doesn't matter. However he does it, if they're only unconscious for a few minutes it probably won't take them long to recover.'

'I didn't realize you were an expert,' Gardner said tartly.

I could have pointed out that I used to be a GP, or that I'd once been drugged myself. But there was no point. Everyone was feeling the strain, and Gardner more than most. No one had emerged from this with any credit, but as the Assistant Special Agent in Charge of the investigation, the final responsibility ultimately lay with him. I didn't want to add to that burden.

Not with Sam's life at stake.

Paul himself seemed to have crossed beyond fear and panic into a state of numb isolation. When he came back from phoning Sam's parents he sat without speaking, staring into the impossible nightmare that had engulfed his life. Her parents would be flying out from Memphis the next day, but he hadn't bothered calling anyone else. The only person he wanted now was Sam; everyone else was an irrelevance.

I felt torn over what to do. I wasn't needed there, but I couldn't simply leave Paul and go back to my hotel. So I sat with him in the lounge as coffee-breathed TBI officers went about their business, and the last hours and minutes of one day ticked inexorably towards the next.

It was just after eleven when Jacobsen came into the lounge. Paul quickly looked up, hope dying in his eyes as she gave a quick shake of her head.

'No news. I just wanted to ask Dr Hunter a couple of things about his statement.'

He sank back into his lethargy as I went out with her. I saw she was carrying a folder in her hand, but it wasn't until we were in the kitchen that she opened it.

'I didn't want to upset Dr Avery with this yet, but I thought you should know. We rechecked the footage from the hospital's security cameras around the time York called Dr Lieberman from the pay-phone. You were right about the ambulance.'

She handed me a black and white photograph from the folder. It was the CCTV still I'd seen before, showing the shadowy figure of York crossing the road by the phone booth. The rear of the parked ambulance was visible at the left hand side of the frame. It was hard to say, but he could have been heading towards it.

'The ambulance arrived ten minutes before York used the pay-phone and left seven minutes later,' Jacobsen said. 'We can't see who was driving, but the timing fits.'

'Why would he have waited ten minutes before making the call?'

'Maybe he had to wait until there was no one around, or perhaps he wanted to savour the moment. Or gather his nerve. Either way, at ten o'clock he went to make the call, then came back out and waited. Dr Lieberman would've been in a hurry, so it should've only taken a few minutes for him to make it outside. When he didn't show York would have waited awhile before realizing something was wrong and getting out of there.'

I played it through in my mind: York glancing anxiously at his watch, his confidence bleeding away when his victim didn't appear. *Just another minute; just one more* . . . And then driving away, furious, to plan his next move.

Jacobsen pulled out another photograph from the folder. This one had been taken in a part of the hospital I didn't recognize. An ambulance was caught in the centre of the frame, blurred by motion.

'This was taken on a different stretch of road a few minutes before the ambulance pulled up outside the mortuary,' she said. 'We traced its route backwards on other security cameras. It's definitely the same vehicle. This is the best shot we've been able to find.'

That wasn't saying much. The photograph had been enlarged to the limits of useful magnification, and had the out-of-focus look of still lifted from video. The angle made it impossible to see whoever was inside the cab, and from what I could see there was nothing remarkable about the ambulance itself: a boxy white van with the predominantly orange livery of East Tennessee's main emergency service.

'How can you be sure this is the same one York used?' I asked.

'Because it isn't a real ambulance. The markings look authentic, but only until you compare them to the real thing. Not only that, but it's a model that's at least fifteen years old. That's way too old to be still in use.'

I examined the photograph more carefully. Now she'd mentioned it the ambulance did look dated, but it was good enough to fool most people. Even in a hospital. Who would think to look twice?

I handed the photograph back. 'It looks pretty convincing.'

'There are companies that specialize in selling used ambulances. York could've probably picked up an old model like this for next to nothing, and then repainted it in the right colours.'

'So can you trace where it came from?'

'Eventually, but I'm not sure how much good that'll do us. York probably used a credit card from one of his victims to buy it. And even if not I doubt it'll help us find him now. He's too smart for that.'

'What about the registration?' I asked.

'We're still working on it. The plates are visible in some shots, but they're too dirty to make out. Could be intentional, but the vehicle's sides are splashed as well, so it's obviously been driven through mud some time recently.'

I thought about what Josh Talbot had said when he'd identified the dragonfly nymph from the casket. *The body had to have been left close to a pond or lake. Probably right by the water's edge . . . They're not called swamp darners for nothing.*

'At least now we've a better idea of what we're looking for,' Jacobsen went on, putting the photographs back into the folder. 'Even without the registration we can release a description of the ambulance. That'll narrow things down a little, if nothing else.'

But not enough. York had been given plenty of time to reach wherever he was going. Even if he hadn't crossed the state line, there were hundreds of square miles of mountain and forest where he could lose himself.

And Sam.

I looked at Jacobsen and saw the same thought in her eyes. Neither of us spoke, but an understanding passed between us. *Too late.* Inappropriate as it was, I was suddenly conscious of how close we were standing, of the way the scent of her body underlay the light perfume after the long day. The sudden awkwardness between us told me she was aware of it as well.

'I'd better get back to Paul,' I said, moving away.

She nodded, but before either of us could say anything else the kitchen door opened and Gardner walked in. One look at his seamed face was enough to tell me that something had happened.

'Where's Dr Avery?' he asked Jacobsen as though I weren't there.

'In the lounge.'

Without a word he went out again. Jacobsen went with him, all emotion carefully smoothed from her face. The air seemed suddenly cold as I followed.

Paul didn't seem to have moved since I'd left him. He still sat hunched in the chair, a mug of coffee standing cold and untouched on the low table beside him. When he saw Gardner he stiffened, holding himself like a man preparing for a physical blow.

'Have you found her?'

Gardner quickly shook his head. 'Not yet. But we've had a report of an accident involving an ambulance on Highway 321, a few miles east of Townsend.' I knew the place by name, a small, pretty town in the foothills of the mountains. Gardner hesitated. 'It isn't confirmed yet, but we think it was York.'

'Accident? What sort of accident?'

'It was in a collision with a car. The driver says the ambulance took a bend too fast and sideswiped him coming the other way. Spun both of them round, and the ambulance went into a tree.'

'Oh, Christ!'

'It took off again, but according to the car driver the front fender and at least one of the lights were smashed. By the sound it made he thinks there could have been some mechanical damage as well.'

'Did he get the registration?' I asked.

'No, but a banged-up ambulance is likely to get noticed. And at least now we know which way York was headed.'

Paul had jumped up from his seat. 'So now you can put up road-blocks?'

Gardner looked uncomfortable. 'It isn't that simple.'

'Why the hell not? For Christ's sake, how hard can it be to find a beat-up ambulance when you know which damn way it's heading?'

'Because the accident was five hours ago.'

There was silence as his words sank in.

'The driver didn't report it straight away,' Gardner went on. 'Seems like he thought it was a real ambulance, and was worried he might get into trouble. It was only when his wife convinced him to try for compensation that he called the police.'

Paul was staring at him. 'Five *hours*?' He sat down, as though his legs would no longer support him.

'It's still a valuable lead,' Gardner insisted, but Paul wasn't listening.

'He's gone, hasn't he?' His voice was flat and lifeless. 'He could be anywhere. Sam could be already dead.'

No one contradicted him. He stared at Gardner with such intensity that even the TBI agent seemed to flinch.

'Promise me you'll catch him. Don't let the bastard get away with this. Promise me that much, at least.'

Gardner looked trapped. 'I'll do my best.'

But I noticed he didn't look Paul in the eye as he said it.

21

They found the ambulance next morning. I'd spent most of that night in an armchair, dozing fitfully. It seemed endless. Each time I woke I'd check my watch and find that only a few minutes had passed. When I looked out of the window and saw a golden glow breaking in the sky, it felt as though time was starting up again.

Glancing at the other armchair, I saw that Paul was wide awake. He didn't seem to have moved all night. I stood up stiffly.

'Do you want a coffee?'

He shook his head. Flexing my neck and shoulders, I went into the kitchen. The coffee had been warming all night, filling the room with a stale, burnt odour. I poured it down the sink and made a fresh pot. I switched off the light and went to stand by the window. Outside, the world was starting to take form in the early morning gloom. Beyond the houses opposite I could just make out the lake, its dark surface smudged with white mist. It would have been a peaceful early morning scene, if not for the patrol car parked outside, a lurid splash of reality in the tranquil dawn.

I sipped my coffee as I stood by the kitchen window. Outside, a bird began to sing. Its lone voice was soon joined by others, a growing chorus of birdsong. I thought about Jacobsen's grim forecast: *If he hasn't killed her already, she'll be dead before the night's out.* As though on cue, the first shafts of sunlight touched the lake.

It was going to be a beautiful morning.

By eight o'clock the first TV crews and reporters began to arrive. Sam's name hadn't been released to the press, but it was always only going to be a matter of time before it leaked out. The uniformed officers stationed in the patrol car made sure the press stayed off the property, but in no time at all the road was choked with news crews and vehicles.

Paul barely seemed to notice. In the daylight he looked awful, the skin on his face grey and lined. He seemed increasingly withdrawn, lost in a private zone of suffering. The only time he came to life was when the phone rang. Each time he would snatch it up, tense with expectation, only to sag a moment later when it was just another friend or persistent journalist. After saying a few words, he'd hang up and retreat back into his shell. I felt for him, knowing all too well what he was going through.

But there was nothing I could do to help.

It was just before noon when the pattern was broken. The remains of sandwiches lay curling on plates beside us. Mine were half eaten, Paul's untouched. I was beginning to think it was time for me to go back to my hotel. I was doing no good here, and Sam's parents would be arriving in a few hours. When the phone rang again Paul grabbed for it, but I could see from the way his shoulders slumped that it wasn't Gardner.

'Hi, Mary. No, I haven't—' He broke off, his entire posture radiating a new urgency. 'What channel?'

Letting the phone drop he grabbed for the TV remote.

'What is it?' I asked.

I don't think he heard. He was flicking through the channels as soon as the screen came to life, scrolling through a cacophony of noise and images until he suddenly stopped. A young woman with lacquered hair and too-red lipstick was talking animatedly to camera.

'. . . *breaking news story, a report is coming in that an ambulance has been found abandoned in the Gatlinburg area of the Great Smoky Mountains National Park . . .*'

Paul's face had gone slack as the impact of the words struck him.

'. . . *exact location has not been revealed, and TBI sources are refusing to confirm that this is the same vehicle used in yesterday's abduction of Samantha Avery, the pregnant thirty-two-year-old from Blount County. There's no word yet on the whereabouts of the missing woman, but unconfirmed reports say the ambulance may have been damaged in a collision . . .*'

The newsreader continued, her voice breathy and excited, as a photograph of York appeared on the screen, but Paul was already grabbing for his phone. It rang before he could dial a number. *Gardner*, I thought, and saw my guess confirmed in Paul's expression.

'Have you found her?' he demanded.

I watched him slowly deflate at Gardner's answer. In the silence I could hear the TBI agent's voice, tinny and indistinct. Paul listened, his face tortured and intent.

'And you let me hear about it on *TV*? For God's *sake*, you said you'd call when you had any news . . . I don't care, just call me, OK?'

He hung up. He stood with his back to me, bringing himself under control before he spoke.

'They found the ambulance half an hour ago at a picnic spot close to I-40,' he said dully. 'They think York abandoned it and stole a car before he got to the Interstate. That'd take him halfway across North Carolina. Unless he headed west. He could be on his way to New Mexico by now. He could be *anywhere!*'

The phone shattered as he hurled it against the wall, scattering plastic across the room.

'Jesus *Christ*, I can't stand this! What am I supposed to do? Just *sit* here?'

'Paul—'

But he was already heading for the door. I hurried after him into the hallway.

'Where are you going?'

'To see the ambulance.'

'Wait a second. Gardner—'

'Screw Gardner!' He started to open the front door. I put my hand on it. 'Get out of my way, David!'

'Just *listen*, will you? If you go out now you'll have TV crews trailing you all the way there. Is that what you want?'

That stopped him.

'Is there a road at the back?' I went on quickly, while I had his attention.

'This one loops around the houses there, but I can't—'

'I'll get my car. The press won't follow me, but it'll distract them. You go out the back and cut across the gardens, and I'll meet you round there.'

He didn't want to, but he could see the sense of what I was saying. Reluctantly, he nodded.

'Give me a couple of minutes,' I told him, and went out before he could change his mind.

Sunlight slapped my face when I stepped outside, dazzling me. I made straight for the car, trying to ignore the sudden clamour my appearance had sparked. The press surged forward, a wall of cameras and microphones, but their excitement was short-lived. 'That isn't Avery,' someone said, and it was as though someone had flicked a switch. A few half-hearted questions were fired at me, but interest quickly waned when I didn't answer. The attention of the TV crews and reporters was already back on the house as I climbed into the car and drove away.

The road meandered round in a slow curve before doubling back on itself behind Sam and Paul's house. The street here was empty, except for Paul. He ran over as I pulled up, and had the door open before the car had even stopped.

'Go back to the main highway and head for the mountains,' he said, out of breath.

No trailing press cars followed us as we left the development. The route was signposted once we reached the highway. Apart from the occasional terse direction from Paul, we drove in silence. The mist-shrouded Smoky Mountains loomed up on the horizon ahead of us. The sight of them stretching into the distance was sobering, bringing home the impossible scale of any search.

The sun was high overhead, warm enough to pass for a summer's day. After a few miles I had to use the screen wash to clear the glass of dead insects. The tension in the car grew as we reached the foothills and drove through Townsend. It wasn't far from here where York had clipped the car and hit a tree. A few miles past the town we came to a tall oak by the roadside that had been ringed with police tape. The jagged white gouges in its bark were clearly visible. Paul stared at it as we drove past, his face bleak.

Neither of us spoke.

A few miles further on, he directed me to branch off the highway and we began to climb into the mountains. They rose up around us, plunging the road in and out of shadow as it wound through them. We saw a few other cars but it was still too early in the season for there to be many. Spring was everywhere. The woods were carpeted with wild flowers, blue, yellow and white dappling the vibrant new grass. At any other time the Appalachian beauty would have been breathtaking; now it seemed like a cruel joke.

'Take the next right,' Paul told me. The turn-off was a narrow road, gravelled like many of the minor roads and tracks out here. This one was steep enough to have the car's automatic transmission straining. After a half-mile it levelled out. We rounded a bend and found our way blocked by a patrol car. Beyond it, I could make out wooden picnic tables and parked police vehicles before trees blocked the view.

I wound down the window as a uniformed deputy approached the car. He looked barely out of his teens, but walked with an older man's swagger. He stared down at me from under the wide brim of his hat, one hand on his holstered gun.

'Back up. Y'all cain't come up here.'

'Can you tell Dan Gardner that Dr Hunter and—' I began, and then I heard the passenger door open. I looked round to see Paul climbing out of the car. *Oh, Christ*, I thought, as the young deputy scrambled to head him off.

'Hold it right there! Goddammit, I said stop!'

I hurried out of the car after them, grabbing hold of Paul as the deputy planted himself on the track in front of him and drew his sidearm. I'd never realized how much I disliked guns until then.

'OK, it's OK,' I said, pulling Paul back. 'Come on, take it easy!'

'Back in the car! Now!' the deputy yelled. He gripped the gun in both hands, pointing it at the ground between us.

Paul showed no inclination to move. In the bright sun his eyes didn't look fully focused. He couldn't touch York, but the need for confrontation was consuming him. I don't know what might have happened, but at that moment a familiar voice rang out.

'What the hell's going on?'

I never thought I'd be glad to see Gardner. The TBI agent was striding down the track, tight-lipped. The deputy glared at Paul, gun still outstretched.

'Sir, I told them they cain't come up here, but they won't—'

'It's all right,' Gardner said, but without enthusiasm. His suit looked more crumpled than ever. He spared me a cold glance before addressing Paul. 'What're you doing here?'

'I want to see the ambulance.'

It was said in the inflectionless tone of someone whose mind is made up. Gardner considered him for a moment, then sighed.

'It's this way.'

We followed him back up the track. The picnic area was set on a grassy clearing overlooking the foothills. They spread out below us, miles of tree-covered peaks and troughs: a frozen ocean of green. This high up the air was cooler but still warm, sweet with pine and spruce.

At one side of the clearing the police vehicles were clustered in front of a handful of civilian cars.

Parked slightly away from them, quarantined by crime tape, was the ambulance.

Even from a distance I could see the damage caused by the collision. Parallel gouges ran along one side, and the left wing had crumpled like tinfoil where it must have hit the tree. Small wonder it had been abandoned; York had been lucky to get as far as he had.

Paul stopped at the police tape and stared into the back of the ambulance. Its doors hung wide open, revealing shabby bunks and cabinets. A forensic agent was busy inside, and we could see restraining straps dangling from one of the bunks, as though they'd been hurriedly flung off.

I felt someone beside me, and turned to find Jacobsen. She gave me a solemn look. There were dark smudges under her eyes, and I guessed Paul and I weren't the only ones who had gone without sleep.

Paul's face was a mask. 'What have you found?'

He didn't seem to notice Gardner's slight hesitation. 'There were blond hairs on the bunk. We'll need to check them against samples of your wife's hair, but we don't think there's much doubt. And it looks like York must have taken quite a knock in the collision.'

He led us round to the front. The driver's door was hanging so we could see into the grubby and well-worn interior. The steering wheel was buckled and skewed slightly to one side.

'Chances are York's pretty banged up himself if he smacked the wheel hard enough to do that,' Gardner said. 'Must've busted a rib or two, at least.'

For the first time something like hope showed on Paul's face. 'So he's injured? That's good, isn't it?'

'Maybe.' Gardner was non-committal.

Something in his tone sounded off, but again Paul was too preoccupied to notice. 'I'd like to stay here for a while.'

'Five minutes. Then you need to go on home.'

Leaving Paul there, I walked away with Gardner and Jacobsen. I waited until we were out of earshot.

'What aren't you telling him?'

Gardner's mouth compressed, but whatever he might have said went unspoken as someone called him from the crime scene truck.

'You might as well let him know,' he told Jacobsen before walking away, the line of his back as uncompromising as ever.

The shadows under Jacobsen's eyes added to her solemnity. 'There are bloodstains in the ambulance. On the bunk and on the floor.'

I pictured Sam as I'd last seen her. *Oh, dear God.* 'Don't you think Paul's got a right to know?'

'Eventually, yes. But not all of the stains are fresh, and we can't say for sure that any of them belong to his wife.' Her gaze flicked to where Paul maintained his vigil by the ambulance. 'Dan doesn't think knowing about it is going to help him right now.'

I reluctantly accepted that. I didn't like keeping information from Paul, but his imagination would be torturing him enough already.

'How did you find the ambulance?' I asked.

She brushed back a strand of hair that had strayed over her face. 'We got a report of a stolen car, a blue Chrysler SUV. There are rental cabins about a quarter of a mile away but they don't have a road. Tenants leave their cars here and hike up the rest of the way. That's probably why York chose this place – even this early in the season there are usually one or two cabins rented out. Anyone familiar with this area would know there'd be cars here.'

I looked over at the damaged ambulance. It had been left out in the open, a few yards from a thick clump of laurel bushes. 'York didn't make much effort to cover his tracks.'

'He didn't have to. Cars can be left here for days while their owners play at pioneers. York could bank on the one he took not being missed till this morning at least, and maybe even longer. It was pure luck that the owner noticed when he did.'

Luck. It wasn't something we'd had much of so far. 'I'd have thought he'd at least have parked it so the damage was less obvious.'

Jacobsen gave a tired shrug. 'I expect he had more important things to think about. He'd got to get Samantha Avery into the car, and that can't have been easy if he was injured himself. Hiding the ambulance would have been the least of his problems.'

That made sense, I supposed. York only needed it to remain undiscovered long enough to get where he was going. After that it wouldn't matter.

'You think he was heading for the Interstate?' I asked.

'That's how it looks. It's only a few miles away, and from there he could go deeper into the mountains, double back west or head for another state.'

'So he could be anywhere.'

'Pretty much, yes.' Her chin came up. She looked over towards the ambulance where Paul was standing. 'You should take him home. This isn't doing anyone any good.'

'He shouldn't have had to find out about it from the TV.'

She nodded, accepting the implied rebuke. 'Dan was going to call him as soon as he had time. But we'll let Dr Avery know straight away if there's any more news.'

I noticed she said *if*, not *when*. The longer this went on the less chance there was of finding Sam.

Not unless York wanted us to.

I went back to Paul as Jacobsen joined Gardner at the crime scene truck. He cut a forlorn figure by the ambulance, staring at it as though it might help him divine the whereabouts of his wife.

'We should go now,' I told him gently.

All the fight he'd shown earlier seemed to have burned out of him. He looked at the ambulance for a second or two longer, then turned his back on it and walked with me to the car.

The young deputy gave Paul a hard stare as we passed him on the track, but it was wasted. Paul didn't seem aware of anything as we left the picnic area behind. We'd gone several miles before he spoke.

'I've lost her, haven't I?'

I searched for something to say. 'You don't know that.'

'Yes I do. So do you. So did everyone back there.' The words were spilling out of him now like water from an overfull cup. 'I keep trying to remember what I said to her last. But I can't. I've been going over and over it in my mind, and there's nothing there. I know it shouldn't bother me, but it does. I just can't believe the last time I saw her was so ordinary. How can I not have known?'

Because you never do. But I didn't say that.

He lapsed into silence. I stared numbly at the road ahead. *Dear Christ, don't let this happen.* But it already had, and the silent woods offered no relief. Insects bobbed through the broken columns of sunlight, insignificant specks beside giant oak and pines that had stood here since long before I was born. A slender waterfall tumbled through a cleft in the hillside, foaming white over dark rocks. We passed fallen trunks covered in moss, others being slowly choked by vines while they still stood. For all its beauty, everything that lived out here was in a constant fight to survive.

Not everything succeeded.

I'm not sure when I became aware of my unease. It seemed to come from nowhere, announcing itself first as a prickling on my forearms. I looked down and saw the hairs on them were standing up; a similar tickling told me those on the back of my neck had started to rise as well.

As if only waiting for that, the disquiet bloomed into a clamouring sense of urgency. I gripped the steering wheel. *What? What's wrong?* I didn't know. Beside me Paul still sat in haunted silence. The road ahead was clear and empty, dappled with sunlight and shadows from the trees. I checked the rear-view mirror. There was nothing to see. Behind us the woods unrolled with indifferent monotony. But the feeling persisted. I glanced in the mirror again, and jumped as something hit the windscreen in front of me with a dull *slap*.

A large insect was mashed against the glass in a tangle of legs and wings. I stared at it, feeling the urgency begin to coalesce. Without thinking what I was doing, I stamped hard on the brake.

Paul braced himself against the dashboard as he was flung against his seatbelt. He stared at me in bewilderment as the car screeched to a halt.

'Jesus, David!' He looked round, trying to see why we'd stopped. 'What's wrong?'

I didn't answer. I sat gripping the steering wheel, my heart bumping against my ribs. I was still staring at the windscreen. The dragonfly was big, almost as long as my finger. It was badly mangled, but I could still make out the tiger-striped thoracic markings. Its eyes were unmistakable, just as Josh Talbot had said.

The electric blue of *Epiaeschna heros*.

A swamp darner.

22

Paul was looking at me as though I'd gone mad as I put the car into reverse.

'What is it? What have you seen?'

'I'm not sure.'

I twisted round in my seat to look through the rear window, scanning the woods on my side as I backed up the road. Talbot had said swamp darners liked wet, wooded habitats. And amongst all the insects, there had been a blue sparkle in the trees I'd been too distracted to notice. Not consciously, at least. *Just look at those eyes! Incredible, aren't they? On a sunny day you can spot them a mile away.*

He'd been right.

I pulled over on to the bank beside the road. Leaving the engine running, I got out and went to stand on the edge of the woods. A green, outdoor silence enveloped me. Sunlight shafted down between the tree trunks and branches, picking out mats of wild flowers growing through the grass.

I saw nothing.

'David, for God's sake will you tell me what's going on?'

Paul was standing by the open passenger door. The sour taste of anticlimax was in my mouth. 'That's a swamp darner on the windscreen. The same as the nymph we found in Harper's casket. I thought . . .'

I tailed off, embarrassed. *I thought I might have seen more of them.* It seemed far-fetched now.

'Sorry,' I said, and turned to go back to the car.

And saw a glint of blue among the green.

'There.' I pointed, my heart thudding. 'By the fallen pine.'

The dragonfly zigzagged through patches of dappled sunlight, blue eyes shining like neon. As though they'd chosen that moment to appear, now I picked out others amongst the trees.

'I see them.' Paul was staring into the woods, blinking as though just waking up. 'You think it's important?'

There was a tentative, almost pleading note to his voice, and I hated myself for raising his hopes. Swamp darners or not, York wouldn't have left Noah Harper's body so close to a road. And even if he had, I couldn't see how it would help Sam. Yet we knew York had headed out this way in the ambulance, and now here were the dragonflies as well. That couldn't just be coincidence.

Could it?

'Talbot said they like standing water, didn't he?' Paul said, with an excitement born of desperation. 'There must be some round here somewhere, a lake or pond. Do you have a map in the car?'

'Not of the mountains.'

He ran his hands through his hair. 'There's got to be something! Perhaps a slow moving creek or stream . . .'

I was beginning to wish I'd not said anything. The mountains covered over half a million acres of wilderness. The dragonflies could be migrating, for all I knew; might already be miles from wherever they'd hatched.

Still . . .

I looked round. A little further down the road I could see what looked like a turning on to a track.

'Why don't we take a look down there?' I said.

Paul nodded, eager to seize even the slimmest hope. I felt another stab of guilt, knowing we were probably just clutching at straws. As he got back into the car I picked the dead dragonfly from the windscreen. When I turned on the wipers the water jets sluiced the remains from the glass, and it was as though it had never been there.

The turning was little more than a dirt track running off through the trees. It didn't even merit a layer of gravel, and I had to slow to a crawl along the rutted, muddy surface. Branches and shrubs scratched at the windows. They grew thicker with every yard, until eventually I was forced to stop. The way ahead was completely blocked, maples and birches fighting for space with straggly laurel bushes. Wherever the track might have once led, we weren't going any further.

Paul banged the dashboard with frustration. 'Goddammit!'

He climbed out of the car. I did the same, forcing open the door against the push of branches. I looked round, hoping to catch a glimpse of another swamp darner, anything that would tell me this wasn't a waste of time. But the woods were mockingly empty.

Paul's shoulders had slumped in defeat as he contemplated the enclosing tangle of trees. The hope that had briefly fired him up had burned itself out.

'This is useless,' he said, his face a carving of despair. 'We're miles from where York abandoned the ambulance. Hell, we're almost back at where he had the accident. We're wasting our time.'

I almost gave up then. Almost got back in the car, accepted that I'd over-reacted. But Tom's words came back to me: *You've got good instincts, David. You should learn to trust them more.*

For all my doubts, my instincts still told me this was important.

'Just give me a minute.'

The branches overhead whispered as a breeze disturbed them, then fell silent again. I went to where a rotting tree trunk sprouted pale, plate-like fungi, and climbed on to it. The vantage point made little difference. Except for the overgrown track we'd followed there was nothing to see but trees. I was about to get down again when the branches overhead stirred and rustled as the breeze returned.

And then I caught it.

The faint, almost sweet taint of decomposing flesh.

I turned my face to the breeze. 'Can you . . .'

'I smell it.'

There was a tightness to his voice. It was an odour both of us were too familiar with to mistake. Then the breeze died, and the air held just the normal scents of the forest.

Paul looked round frantically. 'Did you get where it was coming from?'

I pointed across the hillside, in the direction the breeze seemed to have been blowing from. 'I think it was that way.'

Without a word, he strode off through the woods. I gave the car a last glance, then left it and hurried after him. The going was difficult. There was no path or trail, and neither of us was dressed for hiking. Branches plucked at us as we picked our way along the uneven ground, the thickets of bushes making it impossible to keep to a straight line. For a while we were able to use the car to keep our bearings, but once that was out of sight we had to rely on guesswork.

'If we go much further we're going to get lost,' I panted, when Paul stopped to disentangle his jacket from a low branch. 'There's no point just wandering about without knowing where we're going.'

He scanned the trees around us, chest rising and falling as he gnawed at his lip. Desperate as he was for anything that would lead him to York and Sam, he knew as well as I did that it might just have been carrion we'd caught wind of.

But before either of us could say anything else the branches around us shivered as the breeze picked up. We exchanged a look as we caught the odour again, stronger than ever.

If it was carrion it was something big.

Paul picked up a handful of pine needles and tossed them into the air, watching which way they were blown. 'That way.'

273

We set off again, with more confidence this time. The smell of decay was noticeable even when the breeze dropped now. *You don't need a detector to smell this, Tom.* As though to confirm we were heading in the right direction I caught a metallic shimmer as a dragonfly flashed through the trees up ahead.

Then we saw the fence.

It was partly hidden by scrub pine and bushes, eight feet high wooden slats topped with razor wire. The slats were rotten, and what looked like a much older chain-link fence ran round the outside, rusted and sagging.

Paul seemed charged with an almost feverish energy as we began to pick our way along the boundary. A little further along a pair of old stone gateposts had been incorporated into the fence, now blocked off with wooden slats. The ground in front was overgrown, but deep parallel grooves were still visible.

'Wheel ruts,' Paul said. 'If there're gateposts there must've been a road of some sort. Could be the same track we were following.'

If it was it hadn't been used in a long time.

The smell of decomposition was much stronger now, but neither of us made any comment. There was no need. Paul stepped over the sagging chain-link fence and took hold of one of the wooden slats. There was a splintering *crack* as the rotten wood came away in his hands.

'Wait, we need to tell Gardner,' I said, reaching for my phone.

'And say what?' He wrenched at the fence, grunting with exertion. 'You think he's going to drop everything and come running because we smelled something dead?'

He kicked at a slat until it broke, then began furiously working at another, prying it loose from a stubborn nail with a loud creak. Bushes poked through the gap from the other side, obscuring whatever else might be through there. Tearing away the last few splinters of wood, he spared me a brief glance.

'You don't have to come with me.'

He began to climb through the fence. Within seconds there were just waving branches to show where he'd been.

I hesitated. No one knew where we were, and God only knew what lay behind the fence. But I couldn't let Paul do this alone.

I squeezed through the gap after him.

My heart jumped as something caught hold of my jacket. I tugged at it in a panic until I saw I'd only snagged it on a nail. I pulled free and carried on. Bushes crowded right up to the fence on this side. Ahead of me I could hear snapping and rustling as Paul forced his way through them. I followed as best I could, shielding my face with one hand as twigs clawed at my eyes.

Then I stepped clear and almost walked into him.

We'd emerged into a large garden. Or, rather, what had once been a garden; now it was a wilderness in its own right. Ornamental shrubs and trees had run riot, crowding each other in the fight for space. We stood in the shade of a huge magnolia, the scent from its waxy white flowers cloying and sweet. Directly ahead of us stood an old laburnum, heavy branches dripping with clusters of yellow.

Underneath it was a pond.

It must have once been the garden's centrepiece, but now it was stagnant and rank. Its edges were slowly drying out and choked with reeds, while the viscous green water was filmed with scum. A cloud of midge-like insects danced above its surface like dust motes in the sunlight.

Feeding on them were the dragonflies.

There were dozens of them. Hundreds. The air hummed with their wings. Here and there I saw the iridescent colours of other, smaller species, but it was the tiger-striped swamp darners who ruled, eyes shining like sapphires as they darted in an intricate ballet above the water.

I shifted to get a better view and felt something snap under my foot. Glancing down I saw a pale, green-white stick in the grass. No, two sticks, I thought. And then, like a picture coming into focus, what I was seeing resolved itself into the twin bones of a human forearm.

I slowly stepped back. The body lay half hidden in the undergrowth by my feet. It was fully skeletonized, shoots of bright spring grass already growing through the moss-covered bones.

Black female, adolescent: the assessment came automatically. As though it had been waiting for that moment, now the smell of decomposition reasserted itself over the thick scent of magnolia.

Beside me, Paul spoke in a whisper. 'Oh, my God . . .'

I slowly lifted my gaze. The dragonflies weren't the only inhabitants of this place.

The garden was full of corpses.

They were in the grass, under the trees, in the undergrowth. Many were little more than stripped bones lying in the greenery, but some were more recent; leathery intestines and cartilage still host to flies and maggots. No wonder none of York's earlier victims had been found.

He'd created his own body farm.

Paul's voice was unsteady. 'Over there. There's a house.'

Beyond the pond the ground rose into a tree-covered hillside. Towards its top, the angled lines of a roof were visible through the branches. I grabbed hold of Paul's arm as he started towards it.

'What are you doing?'

He pulled free. 'Sam might be in there!'

'I know, but we've got to tell Gardner—'

'So tell him,' he said, breaking into a run.

I swore, the phone held in my hand. Gardner needed to know about this, but I had to stop Paul from doing anything stupid.

I set off after him.

The corpses were everywhere. They seemed to have been left with no pattern or purpose, as though York had simply dumped them here to rot. Dragonflies swooped and hovered as I ran through the garden, indifferent to the death all around. I saw a swamp darner gently fanning its wings as it rested on a skeletal finger, beautiful but alien. When another thrummed close to my head I batted it away in revulsion.

Paul was still ahead of me, heading for the building we'd seen through the branches. Built on the sloping hillside, it rose up like a cliff, a sprawling timber structure three storeys high. I could see now that it was far too big to be a house, more like an old hotel of some sort. It must have been imposing once, but neglect had made it as rotten as the bodies in its grounds. Its foundations had shifted, giving it a skewed, twisted aspect. Holes gaped in the shingle roof, and cobwebbed windows stared sightlessly from the weathered grey face. Leaning against one corner like a drunk was an ancient weeping willow, its branches draped over the walls as though to hide their decay.

Paul had reached a weed-choked terrace that ran along this entire side of the building. I was close behind him now, but not close enough to stop him as he ran to a pair of boarded-up French doors and wrenched on the handles. They didn't open, but the rattle shattered the garden's silence.

I pulled him aside. 'What are you *doing*? Jesus, do you want to get yourself killed?'

But one look at his face gave me the answer: he didn't expect to find Sam alive. And if she wasn't, he didn't care about himself.

Pushing me away, he ran towards the corner of the building where the old willow leaned against the walls. I couldn't let him get too far ahead, but I daren't wait any longer to call Gardner. I dialled as I ran, relieved to see that there was a weak signal even out here. It was more than I'd hoped for, but I swore when the TBI agent's number went straight to voicemail. There was no time to try Jacobsen; Paul had already vanished under the willow's trailing branches. Gasping out the words, I described where we were as best I could, then snapped my phone shut and sprinted after him.

Up close, the building's rot was obvious. Its wooden siding was as soft as balsa, honeycombed with tiny holes. Thinking about the cloud of insects the dragonflies had been feeding on, I remembered what Josh Talbot had said: *Swamp darners are partial to winged termites.*

They'd found a plentiful supply here.

But I'd more pressing concerns just then. Paul was in sight again up ahead, running up an overgrown path along the side of the building. Chest burning, I made an extra effort and hauled him back before he reached the end of it.

'Get off of me!'

A flailing elbow triggered a starburst of light in my eye, but I didn't let go. 'Just *think*, will you! What if he's got a gun?'

He tried to throw me off. 'I don't care!'

I struggled to hold on to him. 'If Sam's still alive we're her only chance! You want to waste it?'

That reached him. The frenzy died in his eyes, and I felt the resistance ebb from him. Still wary, I let him go.

'I'm not waiting till Gardner gets here,' he breathed.

'I know, but we can't just go charging in. If York's in there let's not make it any easier for him.'

I could see that everything in him wanted to tear down the walls until he found Sam, but he knew I was right. Even though York must know we were there by now, he might not realize there were only two of us. God knew, we didn't have much of an advantage, but announcing our approach would lose what little we had.

Moving more cautiously, we went to the end of the path.

We'd obviously come at the building from the back; now we found ourselves at the front. The spring sun was too low to creep above the high roof, casting a deep shadow. Walking into it was like stepping into cold water. Even the trees on this side seemed darker; towering pines and maples rather than the ornamental varieties at the back. Woodland had reclaimed whatever gardens there used to be, branches meeting over the muddy driveway to form a dark, claustrophobic tunnel that disappeared out of sight.

At one side stood a warped timber sign. The lettering had faded to a ghostly blue that hinted at a long-ago optimism: *Breathe Deep! You're at Cedar Heights Spa and Sanitarium!* It looked to date from the 1950s, and judging by its dilapidation it might have been forgotten ever since.

Though not by York.

Several cars were parked haphazardly on the driveway, stolen along with their owners' lives. Most had obviously not been moved in ages, their roofs and windscreens covered with leaf mould and bird droppings, but two were cleaner than the rest. One was a huge black pick-up truck with darkly tinted windows.

The other was a blue Chrysler SUV.

The realization of how York had fooled us rose like bile in my throat. He must have been almost back here when he'd had the accident. So rather than risk the inevitable search coming too close to Cedar Heights, he'd driven miles out of his way before abandoning the ambulance.

Then he'd stolen a car and doubled back.

The SUV was parked at the bottom of crazy-paved stone steps that led to a roofed veranda. At the top was a pair of tall double doors that had once been grand, but were now as dilapidated as everything else.

One of them stood open.

Paul bent and picked up a wooden strut that had come loose from the veranda as we went up the steps. Through the open door at the top I could make out a large, shadowed foyer and the bottom of a wide staircase. Paul reached out to push the door all the way open.

And my phone rang.

It sounded shockingly loud. I grabbed it from my pocket and saw Gardner's name in the caller display. *Jesus, not now!* I fumbled to answer it but it took agonizing seconds before the piercing trill was silenced.

Gardner's voice crackled unevenly 'Hunter? Where the hell are you?'

But there was no time to answer. No time for anything, because at that moment there was a cry from deep inside the house. It quickly cut off, but Paul's restraint slipped.

'SAM! HOLD ON, I'M COMING!' he yelled, and barged through the doors.

Oh, Christ. But there was no longer any choice. Ignoring Gardner's angry questions, I ran after Paul into the sanitarium.

You cock your head, listening. They'll be here soon; you only have a few minutes. Adrenalin is tingling through you, but you're over the worst of the shock now, able to function again. When you heard them at the French doors the disbelief was paralysing. You'd thought that leaving the ambulance miles away would've thrown them off, allowed yourself to relax.

You should have known better.

Your first instinct was to run, but that wasn't an option. You forced yourself to calm down, to think*! And gradually the panic subsided enough to let you see what you had to do. You're better than them, remember that. Better than anyone.*

You can still turn this round.

You have to hurry, though. The eyes stare at you from the bound figure, wide and terrified, as you make sure the gag won't come out again. You don't want any more screams to tell them where you are, not yet. A sense of waste rises up in you as you start. This isn't how it was meant to be, not when you'd come so close . . . *But there's no time for regrets. No time for anything.*

Only what has to be done.

When it's over you regard your handiwork with distaste. The eyes are no longer staring at you, or at anything else. Your breath comes in ragged bursts as you listen to the sounds of the intruders getting closer. Well, let them. You're almost through. Only one more thing left to do, and then your surprise'll be ready.

Wiping the sweat from your face, you reach for the knife.

23

Paul ran across the foyer. 'SAM? *SAM!*'

His shout bounced off the bare walls. The interior of the sanitarium was dark and empty, stripped of furniture and fittings. The windows were shuttered, letting in only slats and cracks of light. I had an impression of space, of dilapidation and dust, as I plunged after him, the phone clutched to my ear.

'Talk to me, Hunter! What's going on?' Gardner demanded, his words fading in and out as the reception wavered.

'We've found York,' I panted. 'It's an old sanitarium in the foothills, about fifteen, twenty miles from where he left the ambulance. There's . . .' But I didn't know how to describe the nightmare of the garden. I started giving directions to where we'd left the car until his silence checked me. 'Gardner? *Gardner!*'

The connection had failed. I'd no idea how much he'd heard, or even if he'd heard anything at all, but there was no time to call him back. Paul had stopped in the centre of the foyer.

'SAM! WHERE ARE YOU? *SAM!*'

'*Paul!*' I seized hold of him. He shook me off.

'He already knows we're here! *DON'T YOU, YOU BASTARD?*' he bellowed. '*YOU HEAR ME? I'M COMING FOR YOU, YORK!*'

His challenge went unanswered. Our breathing sounded hollow in the cavernous foyer. Either termites or subsidence had undermined the foundations, causing the entire floor to cant drunkenly to one side like a fairground funhouse. Dust coated every surface like dirty felt. Faded wallpaper hung down in swags, and the banisters had been ripped from the once grand staircase in the centre of the room so that its railings stuck up into empty air like loose teeth. Next to it was an old-fashioned lift that had made its last journey decades before, its metal cage rusted and full of debris. There was a smell of age and damp, of mould and rotting wood. And something else.

Although it was faint, the sweetly foul odour of decomposition was here too.

Paul ran to the staircase, footsteps clomping on the wooden floor. The flight leading to the lower floor had caved in, leaving gaping blackness and rubble. He started to go up, but I stopped him, pointing. While one side of the building looked ready to collapse, on the other was a service door marked *Private*. The dusty parquet tiles between it and the entrance were crisscrossed with footprints and thin tyre tracks that could have been from a bike.

Or a wheelchair.

Clutching the wooden spar in his fist, Paul ran across and threw it open. A dark service corridor stretched in front of us, the only daylight coming from a small window at the far end.

'*SAM!*' he yelled.

The shout died to silence. Several doors ran along the corridor's length. Paul ran down it, flinging them back one by one. They banged against the wall with a sound like gunshots, revealing bare cupboards and storage rooms that held only cobwebs. I followed behind him, until we'd reached the last doorway. He yanked it open, and I blinked at the sudden brightness.

An empty kitchen greeted us.

Afternoon sun slanted through filthy windows, giving the room the murky green light of an aquarium. A camp bed stood in one corner, a sleeping bag rumpled on top of it. By its head were shelves made from breezeblocks and raw planks, bowed under the weight of old books. Congealed pans cluttered a huge wood-burning stove, and two huge sinks overflowed with dirty crockery. Standing in the centre of the room was a scarred pine table. The plates on it had been pushed aside to make way for a first aid kit, from which a length of leftover bandage still trailed. Remembering the buckled steering wheel in the ambulance, I felt a savage satisfaction.

It was only when I looked away from the table that I realized one entire wall was covered in photographs.

York had created a montage of his victims; black and white images of agonized faces, just like those I'd seen at his house. There were too many to take in at once, men and women of all ages and ethnicities, pinned up on the wall like some sick gallery. Some of the photographs had started to curl and yellow with age. Wallets, purses and jewellery had been heaped in an untidy pile on a shelf below them, tossed aside as casually as the lives of their owners.

I felt a sudden, feathery vibration as something sticky brushed against my face. I recoiled, almost knocking over a chair before I realized it was only a strip of flypaper. A swamp darner was caught on it, still alive but hopelessly entangled, its fitful struggles only trapping it more. Other strips hung all over the kitchen, I saw, their surfaces crusted with dead flies and insects. York hadn't bothered to take them down, just hung fresh strips until there was hardly any space left.

Paul crossed to where a long-bladed knife lay by the stove. Picking it up, he wordlessly passed me the strut he'd been carrying. It felt flimsy and rotten, but I still took it.

Two doors led off from the kitchen. Paul tried to open the first, but it had warped in its frame. He threw his shoulder against it and it gave with a splintering crack. Off balance, he staggered inside and collided with the pale body hanging from the ceiling.

'Jesus!'

He stumbled back. But it was only the carcass of a pig, split in half lengthways and suspended by its hind leg from a meat hook. The small cupboard-sized room was an old-fashioned cold locker, but the rank smell and buzzing flies told that it wasn't cold enough. Cuts of meat lay bagged and parcelled on the shelves, and a pig's head sat on a bloodstained platter like a sacrificial offering.

Pig's teeth and blood. York didn't like to waste anything.

Paul stared for a moment, chest rising and falling, then went to the remaining door. This one opened smoothly, and I let out my breath when I saw it only led to a small staircase that descended into shadows.

Then I saw the wheelchair pushed to one side at the top.

It was scuffed and battered, and in the half-light I could make out wet smears on the seat. Remembering what Jacobsen had told me about the bloodstains in the ambulance, I glanced at Paul, hoping he hadn't noticed. But he had.

He took the stairs three at a time.

I went after him, conscious of the creak and sway of the rickety staircase. At the bottom was a dark and narrow corridor. Chinks of light seeped through boarded-up windows and a set of French doors; the same ones we'd tried from the outside, I realized. The sanitarium had been built on the hillside, and now we were on the lower ground floor. The smell of decomposition was stronger down here, even stronger than outside. But the corridor was empty, except for a single door at the far end.

A brass sign on it bore the legend *Spa Rooms*.

Paul had already started towards it when a sudden noise cut through the silence. It was like air escaping from a valve, a high-pitched keening that sounded both inhuman and agonized. It cut off as quickly as it started, but there was no doubt about its source.

It came from the spa.

'*SAM!*' Paul bellowed, and charged for the door.

I couldn't have held him back even if I'd wanted to. Gripping the length of wood so tightly my hand hurt, I was right behind him as he burst through. There was just time to register a large room with white-tiled walls before a figure dashed through another doorway right in front of me.

My heart stuttered until I realized it was my own reflection.

A huge mirror was fixed to the opposite wall, its surface mottled and leprous. A row of drinking fountains stood in front of it, their spigots dusty and dry. A murky light filtered in through a row of high, cobwebbed windows, revealing cracked white tiles from floor to ceiling. Signs proclaiming *Treatment Rooms*, *Sauna*, and *Turkish Bath* pointed off towards the warren of shadowed chambers that led from the room in which we stood. But we barely noticed.

York had left his victims in here as well.

A sunken plunge pool, perhaps six feet square, stood in one corner by a darkened archway. York had turned it into a charnel pit. The bodies nearly filled it. From what I could make out, they were in varying stages of decomposition, but none so far gone as those outside.

The smell was indescribable.

The sight checked Paul, but only briefly. He quickly crossed to the doors marked *Treatment Rooms* and tore open the nearest one. Inside was a small chamber that must once have been used for massage. Now it was York's darkroom. A reek of chemicals greeted us. Developing trays and containers of photographic chemicals cluttered an old desk, and more photographs had been clipped to a length of cord suspended above it.

Pushing past me, Paul ran to the next chamber. The smell told me what was inside, overwhelming even the darkroom's pungent chemicals. I was overcome by a reluctance to look, a sudden fear of what we were going to find. Paul, too, seemed to feel it. He hesitated, his face deathly.

Then he opened the door.

More of York's victims lay on the tiled floor, stacked one on top of another like so much firewood. They were fully clothed, apparently just dragged in here and left, as though he'd simply lost interest and dumped them in the nearest space to hand.

The body lying on the very top might have been asleep. In the dim light from the doorway, the outflung hand and spill of blond hair looked pitifully vulnerable.

I heard Paul give a sound halfway between a sob and a cry.

We'd found Sam.

24

It was as though all the breath had been sucked out of me. Even though I'd told myself Sam was probably already dead, that York had no reason to let her live, I'd not fully accepted it.

I grabbed hold of Paul as he flung himself forward. 'Don't . . .'

I'd seen the photographs of York's victims. Paul didn't need to see Sam like that. He strained against me, but then his legs gave way. He took a faltering step backward and slid down the wall.

'Sam . . . Oh, Christ . . .'

Move, I told myself. *Get him out of here.* He was slumped on the floor like a broken toy. I tried to get him to his feet.

'Come on. We need to go.'

'She was *pregnant*. She wanted a boy. Oh no, God . . .'

My throat ached. But we couldn't stay there, not when we didn't know where York was.

'Get up, Paul. You can't help her now.'

But he was past listening. I would have tried again, but the tiny chamber suddenly darkened. I jerked round, only to find that the door had swung shut behind us. I quickly pushed it open again, half expecting to see York standing outside. No one was there, but as the grey light from the doorway reached Sam's body, I saw something else.

A glint of silver beneath the tangled blond hair.

There was a clenched feeling in my chest as I stepped nearer to the piled bodies. It grew tighter as I gently moved the hair aside. I felt myself sway when I looked down at the familiar face. *Oh, God.*

Behind me I could hear Paul starting to weep.

'Paul . . .'

'I let her down. I should have—'

I gripped his shoulders. 'Listen to me, *it isn't Sam!*'

He lifted his tearstained face.

'It isn't Sam,' I repeated, letting him go. My chest hurt at what I was about to say. 'It's Summer.'

'Summer . . . ?'

I stood back as he climbed to his feet. He approached the body fearfully, as though not quite believing it even now.

But the steel ear and nose studs were enough to convince him it wasn't his wife. He stood with the knife held limply by his side, taking in the bleached blond hair that had tricked us. The student was lying face down, her head turned to one side. Her face was horribly congested, the single bloodshot eye that was visible dull and staring.

I'd assumed Summer hadn't come to the morgue because she was upset over Tom's death. And instead York had been claiming yet another victim.

A tremor ran through Paul. 'Oh, Jesus . . .'

Tears were streaming down his face. I could guess at the turmoil he was feeling: relief, but also guilt. I felt it myself.

He pushed past me out of the chamber.

'*SAM! SAM, WHERE ARE YOU?*'

His shout reverberated off the tiled walls of the spa. I went after him. 'Paul—'

But he was past restraint. He stood in the centre of the spa, the knife clenched in his fist.

'*WHAT HAVE YOU DONE WITH HER, YORK?*' he yelled,

288

his face contorted. '*COME OUT, YOU FUCKING COWARD!*'

There was no answer. Once the echoes had died, the silence seemed to condense around us. The slow *drip, drip* of an unseen tap counted away the moments like a distant pulse.

Then we heard something. It was faint, the merest suggestion of a sound, but unmistakable.

A muffled whimper.

It came from one of the other treatment chambers. Paul ran and flung the door open. Battery-powered storm lanterns had been arranged around the walls, though none were switched on now. But enough light fell through the doorway to see the unmoving figure in its centre.

Paul's knife clattered to the floor. '*Sam!*'

I groped for the nearest lamp and turned it on, blinking in the sudden brightness. Sam was tied to an old massage table. A camera had been positioned on a tripod by her head, its lens pointing directly down at her face. A wooden chair stood next to it, echoing the arrangement we'd found in the mountain cabin. Her wrists and ankles had been secured by broad leather straps, and a thinner one had been fastened round her throat, tight enough now to dig into the soft flesh. It was connected to a complicated arrangement of steel cogs from which a wooden winding handle protruded.

York's Spanish windlass.

All that registered in the first seconds of reaching the small chamber. *You're too late*, I thought, seeing the tautness of the strap circling her neck. Then Paul shifted to one side, and I saw that Sam's eyes were wide and terrified, but alive.

Her swollen belly looked impossibly big as she lay bound to the table. Her face was red and tear-streaked, and a thick rubber gag had been forced into her mouth. She sucked in a gasping breath as Paul took it out, but the strap round her throat restricted her breathing. She tried to speak, chest working as she gasped for air.

'It's all right. I'm here now. Don't move,' Paul told her.

I went to unfasten the straps holding Sam's ankles, and my foot slipped on something wet. I looked down and saw dark splashes pooled on the white floor tiles. Remembering the bloodstains in the ambulance, I felt cold, until I realized the fluid wasn't blood.

Sam's waters had broken.

I tore at the ankle straps with a new urgency. Next to me Paul reached for the windlass handle.

'Don't touch it!' I warned. 'We don't know which way it turns.'

As badly as we needed to get Sam out of there, the windlass strap was already digging into her throat. If we tightened it by mistake it could kill her.

Indecision racked Paul's face. He started casting around on the floor. 'Where's the knife? I can cut—'

An ear-splitting bellow drowned him out. It came from behind us, from beyond the darkened archway by the plunge pool. It rose in pitch, sounding barely human as it reverberated off the walls before dying away.

The distant tap dripped in the silence. Paul and I stared at each other. I could see his mouth frame a question.

Then York lurched through the archway.

The undertaker was barely recognizable. His dark suit was filthy and stained, his hair matted. The cords on his neck stood out as thick as pencils as he screamed at us, brandishing a long-bladed knife in both hands. Even from where I stood I could see the blood on it, staining his hand black in the poor light.

My limbs felt numb and heavy as I grabbed the wooden strut I'd dropped.

'Get her out!' I told Paul, my voice unsteady, and stepped out to face York.

He came towards me at a shambling run, roaring as he slashed the air with wild swipes of the knife. The strut seemed pathetically flimsy in my hands. *Just give them time. Forget everything else.*

'Wait!' I yelled. Or thought I did; afterwards I was no longer sure if I'd actually said it out loud.

'*Drop the knife!*'

The shout came from the corridor leading to the stairs. Relief surged through me as Gardner emerged through the doorway, Jacobsen close behind. Both had their guns drawn, levelling them at York in a two-handed grip.

'Drop the knife! *Now!*' Gardner repeated.

York had turned towards them. His mouth hung open, panting. There was time to think he was going to do it, that this was going to end here.

Then, with an incoherent scream, he lumbered at Jacobsen.

'Stay back!' Gardner yelled.

York yelled something unintelligible but didn't stop. Jacobsen seemed frozen. I could see the pale fixity of her face as he bore down on her with the knife, but she didn't move.

There were two loud *cracks*.

They were deafening in the tiled confines of the room. York seemed to trip. He stumbled sideways, falling into the big wall mirror. It shattered as he collapsed on to a drinking fountain, dragging it to the floor in a cascade of plaster and silver fragments.

The echoes of gunfire and breaking glass slowly died away.

My ears rang painfully. A faint blue mist hung in the air, a bonfire reek of cordite overlying the stink of decomposition. York didn't move. Gardner hurried over. Still pointing the gun at him, he kicked at the hand holding the knife to knock the weapon away, then quickly knelt and felt at York's throat.

Without urgency, he stood up and tucked the gun back into his belt clip.

Jacobsen was still holding her own gun outstretched, although now it was pointing down at the floor.

'I − I'm sorry,' she stammered, as colour rushed back into her cheeks. 'I couldn't . . .'

'Not now,' Gardner said.

There was a sudden sob from the treatment room. I turned to see Paul helping Sam to sit up, trying to calm her as she coughed and gasped for breath. He'd cut the windlass strap, but a livid red line circled her throat like a burn.

'Oh, G-God, I thought . . . I th-thought . . .'

'Shh, you're all right, it's all right, he can't hurt you now.'

'I c-couldn't *stop* him. I told him I was p-pregnant, and he said . . . he said that was *good*, that he wanted to wait until, wait until . . . Oh, *God!*'

She doubled up as a contraction rocked her. 'Is she OK?' Gardner asked.

'She's in labour,' I told him. 'You need to get an ambulance.'

'On its way. We were heading back to Knoxville when I got your message. I put the call in for back-up and paramedics right away. Christ, what the hell were you *thinking*?'

But I'd no time for Gardner's indignation, or to ask how they'd managed to find us so quickly from my garbled directions. Sam's face was screwed up in pain as I went to her.

'Sam, an ambulance is on its way. We're going to get you to a hospital, but I need you to tell me if you've any other wounds or injuries apart from your throat.'

'N-no, I – I don't think so, he just put me in here and *left* me! Oh, my God, all the *bodies* outside, they're all *dead* . . .'

'Don't worry about those. Can you tell me when your contractions started?'

She tried to concentrate as she panted for breath. 'I don't . . . in the ambulance, I think. I thought it was some *mistake* when he came to the door. He said I should call Paul but when I turned my back he . . . he put his arm round my neck and . . . and *squeezed* . . .'

She was describing a chokehold, I realized. Done properly it could cause unconsciousness in a matter of seconds, with no lasting after-effects. Misjudged, it could kill just as easily.

Not that York would have cared about that.

'I couldn't *breathe*!' Sam sobbed. 'Everything went black, and then I woke up in the ambulance with this *pain* . . . Oh, Lord, it *hurts*! I'm going to lose the baby, aren't I?'

'You're not going to lose the baby,' I told her, with more confidence than I felt. 'We're going to get you out of here now, OK? Just sit tight for two more minutes.'

I went out into the spa, pulling the door to the treatment chamber closed behind me. 'How long till the paramedics arrive?' I asked Gardner.

'Out here? Maybe another half-hour.'

That was too long. 'Where's your car?'

'Parked out front.'

That was an unexpected bonus. I'd thought they'd have come across the hillside as Paul and I had, but I was too concerned about Sam to wonder about it for long.

'The sooner we get Sam out of here the better,' I said. 'If we get her to your car we can meet the ambulance on its way.'

'I'll get the wheelchair from upstairs,' Jacobsen offered.

Gardner gave a short nod, and she hurried out. Grim-faced, he considered the corpses in the plunge pool.

'You say there're more outside?'

'And in here.' With a pang of regret, I told him about Summer's body lying in the other treatment chamber.

'God almighty.' Gardner looked shocked. He passed a hand over his face. 'I'd appreciate it if you stayed behind. I need to hear what happened.'

'Who's going to drive them?' Paul was in no fit state, not with Sam as she was.

'Diane can go. She knows the roads better than you do.'

I looked at the corpses lying on the floor of the spa. I didn't want to stay there any longer than I already had. But I'd trained as a GP, not an obstetrician. I knew Sam would be best served by someone who could get her to the ambulance as soon as possible.

If I belonged anywhere, it was here.

'All right,' I said.

Gardner and I stayed by the unbolted French doors after Jacobsen left with Sam and Paul. It had been decided it was better for them to go out that way rather than risking carrying her up the rotting staircase. Gardner had phoned to check on the progress of the back-up and ambulance, then gone to see if there was another way out through the spa. He reported that the rooms beyond the archway were blocked off.

'Explains why York didn't just take off,' he said, dusting off his hands. 'Must've been down here when you came in and couldn't get out without going past you. Looks like half the floor above has collapsed through there. Whole damn place is being eaten by termites.'

Which in turn had attracted the swamp darners. York's own hiding place had given him away in the end. There was a poetic justice there, but I was too tired to spend long thinking about it.

Jacobsen said little before they left. I guessed she was still reproaching herself over her failure to shoot York. Hard as it must have been, for a field agent that sort of hesitation could be disastrous. If nothing else, it would leave a black mark on her record.

If not for Gardner it could have been far worse.

When they'd gone neither he nor I made any move to go back inside. After the shuttered horrors of the spa, emerging into the sunlight was like being reborn. The breeze carried the smell away from us, and the air was sweet with grass and blossom. I breathed deeply, trying to clean the foulness from my lungs. From where we stood, the trees screened what lay in the garden. With the green mountains rolling to the horizon, it was almost possible to think this was a normal spring day.

'Do you want to take a look down there?' I asked, looking down at the pond glinting through the trees.

Gardner considered it without enthusiasm. 'Not yet. Let's wait till the crime scene truck gets here.'

He still showed no inclination to go back inside. He stared down the hillside towards the pond, hands thrust deep into his pockets. I wondered if it was to stop them shaking. He'd just killed a man, and no matter how unavoidable it might have been that couldn't be easy to deal with.

'Are you OK?' I asked.

It was like watching a shutter come down across his face.

'Fine.' He took his hands from his pockets. 'You still haven't told me what the hell you thought you were doing, coming in here by yourselves. Do you have any idea how stupid that was?'

'Sam would be dead if we hadn't.'

That took the heat out of him. He sighed. 'Diane thinks York was waiting till the last minute, right till she was actually giving birth. He would've wanted to make the most of the opportunity. Two lives for one.'

Christ. I stared across at the mountains, trying to dispel the images that had been conjured.

'You think she'll be OK?' Gardner asked.

'I hope so.' Providing they got her to hospital in time. Providing there were no complications with the baby. It was a lot to hope for, but at least now she had some sort of chance. 'How did you manage to get here so fast? I wasn't sure you'd heard my directions.'

'We hadn't. At least, none that made sense,' he said, with a touch of his old acerbity. 'We didn't need to, though. After York left the skin on the windscreen we put a Bird Dog on your car.'

'A what?'

'A GPS tracking device. We knew where you'd left the car, but the old road you took isn't on any maps. So I took the one that seemed nearest and it led us right to the front gate.'

'You put a *tracker* on my car? And didn't bother to tell me?'

'You didn't need to know.'

That explained why I hadn't seen anyone following me the night before, and how the TBI agents had arrived at Paul and Sam's so quickly. I felt a flash of annoyance that no one had seen fit to let me know about it, but under the circumstances I could hardly complain.

I was just glad it had been there.

'So how did you know you'd got the right place?' I asked.

He gave a shrug. 'I didn't. But there was a new padlock on an old gate, so someone obviously wanted to keep people out. We'd bolt cutters in the trunk, so I cut the lock off and came to take a look.'

I raised my eyebrows at that. Breaking into private property without a warrant was a cardinal sin, and Gardner was a stickler for protocol. His face darkened.

'I decided your phone call constituted probable cause.' His chin came up. 'Come on, let's get back inside.'

The cloying odour of decomposition wrapped itself around us as we went back down the corridor. The light from the French doors didn't reach into the spa, and after the bright sunshine the dim chambers seemed more dismal than ever. Even though I knew what to expect, it didn't lessen the impact of seeing the corpses heaped in the plunge pool like so much rubbish.

York's body lay as we'd left it, as unmoving as his victims.

'Lord, how did he stand the *smell*?' Gardner said.

We went into the small chamber where we'd found Sam. The severed ends of the leather strap that Paul had cut from her throat lay like a dead snake on the old massage table. The windlass bolted to its head had been crafted with obvious care. The ends of the strap fed into an intricate arrangement of finely machined cogs, operated by a polished wooden handle. Turning it would cause the strap to tighten, while the cogs would prevent it from slipping when the handle was released.

A much simpler construct would have been just as effective, but that wouldn't have been good enough for York. Narcissist that he was, he wouldn't have been satisfied with a cord twisted round a piece of wood.

This was his life's work.

'Helluva device.' Gardner sounded almost admiring. Suddenly, he stiffened, cocking his head. 'What's that?'

I listened, but the only sound was the still-dripping tap. Gardner was already out of the treatment room, hand poised on his gun. I followed him.

Nothing in the spa had changed. York still lay unmoving, the blood pooled around him as black and still as pitch. Gardner quickly checked through the archway leading to the blocked-off rooms. He relaxed, letting his jacket fall over his gun again.

'Can't have been anything . . .'

He seemed embarrassed, but I didn't blame him for being jumpy. I'd be relieved myself when the back-up arrived.

'You better show me the other bodies,' Gardner said, all business again.

I didn't go with him into the small chamber where Paul and I had found Summer. I'd already seen more than I wanted. I waited in the spa, standing by York's body. It lay sprawled on its side in the shards of broken mirror, the jagged fragments like silver islands in the blood.

I stared down at the unmoving form, struck as ever by the gulf between its utter immobility and the roaring energy it had possessed a short while ago. I felt too empty for either hate or pity. All the lives York had sacrificed had been a futile attempt to answer a single question: *Is this all there is?*

Now he had his answer.

I was about to turn away, but something stopped me. I looked back at York, uncertain whether I was imagining it. I wasn't.

Something was wrong with his eyes.

Careful to avoid the blood, I crouched beside the body. The sightless eyes were so bloodshot that they looked scalded. The skin around them was badly inflamed. So was his mouth. I leaned forward and flinched back as acrid fumes made my own eyes water.

Darkroom chemicals.

My heart was thumping as I tugged York's body on to its back. The bloodstained hand with the knife flopped limply as it rolled over. I remembered how Gardner had kicked at it before checking his pulse, yet the knife remained clenched in the dead fist. Now I saw why.

Clotted with drying blood, York's fingers had been nailed to the handle.

In that instant, everything fell into place. The agonized keening and York's unintelligible screams; the frenzied slashes of the knife. He'd have been in agony, the toxic chemicals searing his mouth and all but blinding him as he'd tried to pull the nails from his hand. We'd seen only what we'd expected, the crazed attack of a madman, but York hadn't been attacking us.

He'd been begging for help.

Oh, dear God. 'Gardner!' I shouted, starting to scramble to my feet.

I heard him emerge from the chamber behind me. 'For Christ's sake, what the hell do you think you're *doing?*'

What happened next unfolded with the treacle-slow helplessness of a dream.

The remains of the big mirror that York had broken was still fixed to the wall in front of me. In its fragmented surface I saw Gardner pass the plunge pool. As he did, one of the bodies in it moved. My voice died as it detached itself from the others and rose up behind him.

Time started up again. I gave a shout of warning, but it came too late. There was a strangled cry, and I came to my feet to see Gardner struggling to pull free of the arm that was clamped vice-like round his throat.

Chokehold, I thought, dumbly. Then the figure standing behind him shifted its grip, and I felt a shock of recognition as the dirty light from the shuttered windows fell on to its face.

Kyle was breathing raggedly through his open mouth. The round features were the same, but this wasn't the amiable young morgue assistant I remembered. His clothes and hair were clotted with fluid

298

from the putrefying bodies, and his face had a deathly, consumptive pallor. But it was his eyes that were the worst. Without the usual smile to disguise them, they had the flat, empty look of something already dead.

'Move and I'll kill him!' he panted, tightening his hold.

Gardner was clawing at the constricting arm, his face congested, but he didn't have the leverage to pry it loose. I felt a surge of hope as he dropped one hand to the gun at his belt. But he was already losing consciousness, his coordination failing as his brain was starved of blood and oxygen. As I watched his hand limply fell away.

Stooping under the agent's dead weight, Kyle jerked his head towards the treatment room where we'd found Sam.

'In there!'

I was still trying to force my mind to work. How long had Gardner said it would be before the first TBI agents arrived. Half an hour? *How long ago was that?* I couldn't remember. Broken pieces of mirror crunched underfoot as I automatically took a step towards the small chamber. Then I saw the massage table, its leather straps open and waiting.

I stopped.

'Get in there! *Now!*' Kyle roared. 'I'll kill him!'

I had to moisten my mouth before I could answer. 'You're going to kill him anyway.'

He stared at me as though I'd spoken a different language. The pallor of his face was even more noticeable now, shockingly white against the black stubble and bruised skin under his eyes. A greasy sheen of sweat filmed his skin like Vaseline. He was wearing what looked like a medic's uniform, although it was so filthy it was hard to tell.

It could easily have passed for a security guard's.

'*Do* it!' Kyle yanked on Gardner's neck, jerking the TBI agent like a doll. I couldn't tell if he was still breathing, but if the pressure was sustained much longer there'd be brain damage even if he survived.

I bent and picked up a piece of broken mirror. It was long and thin, like a knife. Its edges gouged my palm as I gripped it tightly, hoping Kyle wouldn't see my hand shaking.

He watched me uneasily. 'What're you doing?'

'Let him breathe.'

He tried to sneer, but it was as brittle as the shard of mirror. 'Think you can hurt me with that?'

'I don't know,' I admitted. 'But do you want to find out?'

His tongue darted out over his lips. Kyle was a big man, fleshy and heavily built. *Just like York.* If he dropped Gardner and rushed me I doubted I'd have a chance. But his eyes kept going to the glass shard, and I saw the doubt in them.

He slackened the chokehold enough to let Gardner draw a few rattling breaths, then tightened it again. I saw him flick a look at the doorway.

'Just let him go and I promise I won't try to stop you.'

Kyle gave a wheezing laugh. '*Stop* me? You're giving me your *permission?*'

'His back-up's going to be here any second. If you go now you might—'

'And let you tell them who I am? You think I'm stupid?'

He was a lot of things, but not that. *Now what?* I didn't know. But I didn't think he did either. He was sucking in breaths, stooped and flushed with the effort of supporting Gardner's weight. From the corner of my eye I could see the gun on the agent's belt. Kyle obviously hadn't thought of it so far.

If he did . . .

Keep him talking. I gestured towards York's body. 'Did you enjoy it, mutilating him like that?'

'You didn't give me a choice.'

'So he was just a *diversion?* You did that to him just so you could get away?' I didn't have to try to put contempt into my voice. 'And it didn't even work, did it? All that for nothing.'

300

'You think I don't *know* that?' The shout made him wince, as though in pain. He glared at the undertaker's body. 'Jesus Christ, do you have any idea how much *time* I spent on this? How much *planning*? This isn't how it was supposed to be! York was my way out, my happy fucking *ending*! He'd have been found with Avery's wife, some loser who'd committed suicide rather than be caught. End of story! I'd have left Knoxville afterwards, started out somewhere new, and now look! Goddammit, what a *waste*!'

'No one would have believed it.'

'No?' he spat. 'They believed the photographs I left at his house! They believed everything else I wanted them to!'

A pulse had started to beat in my temple at the mention of Sam. 'And if they had, what then? Murder more pregnant women?'

'I wouldn't have had to! Avery's wife was so full of *life*! She was the one. I could *feel* it!'

'Like you could feel it with all the others? Like you did with *Summer*?' I yelled, forgetting myself.

'She was Lieberman's pet!'

'She *liked* you!'

'She liked Irving more!'

That shocked me to silence. We'd all assumed that Irving had been targeted because of the TV interview. But Kyle had been present that day in the morgue when the profiler had flirted with Summer. The next day Irving had gone missing.

And now Summer was lying in the dark as well.

She only smiled back at him. That was all. For Kyle's ego it had obviously been enough.

I felt sick. But Kyle had become distracted enough to relax his grip on Gardner. I saw the TBI agent's eyelids start to twitch open, and said the first thing that came into my mind.

'What had you got against Tom? Was he such a threat?'

'He was a *fraud*!' Kyle's face twisted in a spasm. 'The big forensic anthropologist, the *expert*! Basking in the glory, playing *jazz* while he

301

worked, like he was in some pizza bar! Hicks was just an asshole, but Lieberman thought he was something *special!* The greatest mystery in the universe right under his nose, and he didn't have the imagination to look beyond the rot!'

'Tom knew better than to waste time searching for answers he couldn't find.' I could hear Gardner wheezing again now, but I daren't spare him a glance. 'You don't even know what it is you're looking for, do you? All the people you've killed, these bodies you've . . . you've *hoarded*, and what for? There's no purpose to any of it. You're like a kid prodding something dead with a stick—'

'*Shut up!*' Spittle sprayed from his mouth.

'Do you even *know* how many lives you've wasted?' I shouted. 'And why? So you can take *photographs*? You think that's going to show you anything?'

'*Yes!* The right one can!' His mouth curled. 'You're as bad as Lieberman, you only see the dead meat. But there's more than that! *I'm* more than that! Life's *binary*, it's on or off! I've stared into people's eyes and watched it go out of them, like flicking a switch! So where's it go? Something *happens*, right then, at that moment! I've *seen* it!'

He sounded desperate. And suddenly I realized that's exactly what he was. That was what this was all about. We'd been wrong about the killer's identity, but Jacobsen had been right about everything else. Kyle was obsessed with his own mortality. No, not obsessed, I realized, looking at him.

Terrified.

'How's your hand, Kyle?' I asked. 'I'm guessing you only pretended you'd stabbed it on the needle. Tom thought he was doing you a favour asking you to help Summer, but you were only hanging round hoping to see one of us get stuck, weren't you? What happened, did you lose your nerve?'

'Shut up!'

'The thing is, if you were just pretending, how come you went so

302

white? It was when I asked about your shots, wasn't it? You'd not thought about infections from any of the people you'd killed until then, had you?'

'I told you to shut up!'

'Noah Harper's tested positive for Hepatitis C. Did you know that, Kyle?'

'Liar!'

'It's true. You should have taken up the hospital's offer of post-exposure treatment. Even though you didn't prick yourself on one of the needles, it was still an open wound. And there was all that gore on your glove. But then you weren't planning on staying around, were you? Much easier to stick your head in the sand than accept you might be infected by one of your own victims.'

His face had paled even more. He jerked his head towards the treatment room. 'Last time! Get in there, *now!*'

But I didn't move. Each minute I kept him talking was a minute closer to help arriving. And looking at his pallor, the ragged way he was breathing, I'd started to think about something else. Why had he chosen to hide, gambling everything on being able to slip out while we were distracted with York, instead of making a run for it while he had the chance? *Perhaps for the same reason he hadn't killed Sam. The same reason he hadn't already choked the life out of Gardner and overpowered me.*

Because he couldn't.

'You took quite a knock in the crash, didn't you?' I said, trying to keep my tone conversational. He regarded me with a hunted expression, his chest rising and falling unevenly. 'I saw the steering wheel in the ambulance. Must have given your ribs a hell of a crack. Did you know that's one of the most common causes of death in car crashes? The ribs splinter and pierce the lungs. Or the heart. How many times have you seen injuries like that in the morgue?'

'Shut up.'

'That sharp, stabbing pain you feel every time you draw a breath?

That's the bone splinters lacerating your lung tissue. It's hard to breathe, isn't it? And it's going to get a lot harder, because your lungs are filling up with blood. You're dying, Kyle.'

'*SHUT THE FUCK UP!*' he screamed.

'If you don't believe me, take a look at yourself.' I gestured to the broken mirror on the wall. 'See how pale you are? That's because you're haemorrhaging. If you don't get medical help soon you're going to either bleed to death or drown in your own blood.'

His mouth worked as he stared at his shattered reflection. I'd no idea how badly hurt he really was, but I'd just fed his imagination. To someone as self-obsessed as Kyle that would be enough.

He'd all but forgotten about Gardner. The TBI agent was blinking now as consciousness returned. I thought I saw him shift slightly, as though he were testing the chokehold. *No, not now. Please, just stay still.*

'Give yourself up,' I went on quickly.

'I'm warning you . . .'

'Save yourself, Kyle. If you give yourself up now you can get medical attention.'

He didn't speak for a moment. I realized with a shock he was crying.

'They'll kill me anyway.'

'No, they won't. That's what lawyers are for. And trials take years.'

'I can't go to jail!'

'Would you rather die?'

He was snuffling back tears. I tried to keep the sudden hope from my face as I saw the tension begin to go out of him.

Then Gardner's hand began inching towards his gun.

Kyle saw what he was doing. '*Shit!*' He wrenched hard on Gardner's throat. The agent gave a choked gasp and pawed feebly at his belt as Kyle grabbed with his free hand for the weapon. I lunged towards them, knowing I wasn't going to reach them in time.

There was a sound from the doorway.

Jacobsen stood framed in it, her face blank with shock. Then her hand swept aside her jacket as she went for her own gun.

'*Leave it!*' Kyle yelled, twisting so Gardner was between them.

She stopped, hand resting on the pistol grip. Kyle had Gardner's gun partway out of its clip, but he had to reach at an awkward angle round the agent's body. The silence was broken only by his ragged breathing. Gardner was no longer moving at all. He hung from the chokehold like a sack, his face darker than ever.

Kyle licked his lips, his eyes going to Jacobsen's belt clip.

'Hand away from the gun and let him go!' she said, but for all her authority there was still a quiver to her voice.

Kyle heard it. Adrenalin had given him a new strength. The moon face moved from side to side as he shook his head and smiled. He was back in control. Enjoying himself.

'Oh, I don't think so. I think you need to put *your* gun down.'

'That's not going to happen. Last chance—'

'Shh.' He cocked his head towards Gardner, as though he were listening. 'I can hardly feel your partner's heartbeat. It's getting weaker. Slowing . . . slowing . . .'

'If you kill him there's nothing to stop me shooting you.'

Kyle's smugness vanished. The pink tongue darted out to moisten his lips again, and at that moment there was the thump of footsteps from the floor above. Kyle's eyes widened, and as Jacobsen's attention wavered he snatched the gun from Gardner's belt and fired.

I saw Jacobsen stagger, but she'd already drawn and fired herself. As Kyle let Gardner fall there were two more *cracks* and a section of mirror by my head exploded, spraying me with splinters. Then Kyle's gun clattered to the floor and he dropped as though his strings had been cut.

My ears rang for the second time that afternoon as I rushed to Jacobsen. She was slumped against the doorway, her gun still rigidly levelled at where Kyle lay. Her face was chalk white, in stark contrast to the spreading dark stain on her jacket. It was on her left side, a glistening wet patch between her neck and her shoulder that grew bigger as I looked.

She blinked. 'I'm . . . I think . . .'

'Sit down. Don't try to talk.'

I spared a quick glance at Gardner's unmoving form as I tore open her jacket. I couldn't see if he was breathing, but Jacobsen's situation was more urgent: if the bullet had hit an artery she could bleed out in seconds. Feet were clattering down the stairs and along the corridor but I barely heard. I'd pulled her jacket from her injured shoulder, my breath catching at how her white shirt was soaked with blood, when figures burst through the doorway. Suddenly the chamber was filled with shouting.

'Quick, we need—' I began, and then I was dragged away and thrust face down on to the floor. *Oh, for God's sake!* I started to get up but something struck me roughly between the shoulder blades.

'Stay down!' a voice yelled.

I yelled that there was no time, but no one was listening. All I could see from my vantage point was a confusion of feet.

It seemed an age before I was recognized and let up. Angrily, I shrugged free of the helping hands. People were crouching by Gardner, who had been moved into the recovery position. He was still unconscious, but I could see that at least he was breathing. I turned to where Jacobsen was being attended by two agents. They'd pulled her shirt away from her neck and shoulder on the side where she'd been shot. Her white sports bra was stained crimson. There was so much blood I couldn't see the wound.

'I'm a doctor, let me take a look,' I said, kneeling beside her.

Jacobsen's pupils were dilated with shock. The grey eyes looked young and scared.

'I thought you were talking to Dan . . .'

'It's OK.'

'The . . . the ambulance was only half a mile away, so I came back. Knew something wasn't right . . .' Her voice was slurred with pain.

'York hadn't taken any of the photographs from the house. His parents, all his past. He wouldn't have just left them . . .'

'Don't talk.'

I felt a surge of relief as I saw the blood-filled furrow in her trapezius, the big muscle that runs between neck and shoulder. The bullet had torn a groove across its top, but despite the bleeding there was no serious damage. Another inch or two lower or to her right and it would have been a different story.

But she was still losing blood. I wadded up her shirt and started to apply pressure to the wound when another agent rushed in with a first-aid kit.

'Move,' he told me.

I stood back to give him room. He tore open a sterile gauze pad and pressed it on to the wound hard enough to make Jacobsen gasp, then began expertly taping it into place. He obviously knew what he was doing, so I went over to Gardner. He was still unconscious, which was a bad sign.

'How is he?' I asked the agent kneeling by him.

'Hard to say,' she said. 'Paramedics are on their way, but we weren't expecting to need them. The hell happened here?'

I didn't have the energy to answer. I turned to where Kyle lay sprawled on his back. His chest and stomach were coated with blood, and his eyes gazed sightlessly at the ceiling.

'Don't bother, he's dead,' the agent told me as I reached down to feel his throat.

He wasn't, not quite. There was the faintest whisper of a pulse under the skin. I kept my fingers there, looking down into the open eyes as his heart gave its final stutters. They grew weaker, the gaps between them longer and longer until eventually they stopped altogether.

I stared into his eyes. But if there was anything there I couldn't see it.

'You're hurt.'

307

The agent kneeling by Gardner was looking at my hand. I saw that it was dripping blood. I must have gashed it on the piece of broken mirror, although I'd no memory of it happening. The cut sliced across the existing knife scar on my palm like a thin mouth, blood welling between its lips.

I'd felt nothing until then, but now it started to burn with a cold, clean pain.

I clenched my hand on it. 'I'll live.'

Epilogue

It was raining in London. After the vivid sunshine and lush mountains of Tennessee, England seemed grey and dull. The tube was busy with the tail end of the evening rush hour, the usual day-worn commuters crammed into each other's personal space. I flicked through the newspaper I'd bought at the airport, feeling the usual sense of dislocation as I read about events that had happened while I'd been away. Coming home after a long trip is always like finding yourself transplanted a few weeks in the future, a mundane form of time-travel.

The world had gone on without me.

The taxi driver was a polite Sikh who was content to drive in silence. I stared out at the early evening streets, feeling grubby and jet-lagged after the long flight. My own street looked somehow different when we turned on to it. It took me a moment to realize why. The branches of the lime trees had been barely shading green when I'd left; now they were shaggy with new leaves.

The rain had slowed to a drizzle, varnishing the pavement with a dark gloss as I climbed out and paid the driver. I picked up my flight bag and case and carried them to the front door, flexing my hand slightly when I set them down. I'd taken the dressing off several days before, but my palm was still a little tender.

The sound of the key turning in the lock echoed in the small hall-way. I'd put a stop on my post before I'd gone away, but there was still a forlorn pile of fliers and leaflets on the black and white floor tiles. I pushed them aside with my foot as I carried the cases inside and shut the door behind me.

The flat looked exactly the same as when I'd left it, except dulled by several weeks' accumulation of dust. I paused in the doorway for a moment, feeling the familiar pang of its emptiness. But not so sharply as I'd expected.

I dumped the case on the floor and set my flight bag on the table, cursing as a heavy *clunk* reminded me what was inside. I unzipped the bag, expecting to be greeted by the reek of spilt alcohol, but nothing was broken. I set the odd-shaped bottle on the table, the tiny horse and jockey perched on the cork still frozen in mid-gallop. I was tempted to open it now, but it was still early. Something to look forward to later.

I went into the kitchen. There was a slight chill in the flat, remind-ing me that, spring or not, I was back in England. I switched the central heating back on, then as an afterthought filled the kettle.

It had been weeks since I'd had a cup of tea.

The message icon on my phone was flashing. There were over two dozen messages. I automatically reached out to play them, then changed my mind. Anyone who needed to contact me urgently would have called my mobile.

Besides, none of them would be from Jenny.

I made myself a mug of tea and took it to the dining table. There was an empty fruit bowl in its centre, a slip of paper lying in it. I picked it up and saw it was a note I'd made before I'd left: *Confirm arrival time w. Tom.*

I balled it up and dropped it back in the bowl.

Already, I could feel my old life starting to reclaim me. Tennessee seemed like an age ago, the memory of the sunlit garden of dragonflies and corpses, and the nightmare scenes in the sanitarium, starting to assume the unreal quality of a dream. But it had been real enough.

Forty-one bodies had been recovered at Cedar Heights; twenty-seven from the grounds, the rest from the spa and treatment rooms. Kyle hadn't discriminated. His victims were a random mix of age, sex and ethnicity. Some of them had been dead for almost ten years, and the task of identifying them was still going on. The wallets and credit cards he'd saved speeded the process to an extent, but it soon became apparent that there were more bodies than there were IDs. Many of his victims had been vagrants and prostitutes whose disappearances weren't always noticed, let alone reported.

If Kyle hadn't felt the need to prove himself, he could have carried on indefinitely.

But not all the victims were anonymous. Irving's body had been recovered from the same chamber as Summer's, and amongst the others who had been identified three names stood out. One was Dwight Chambers. His wallet and driver's licence were in the pile in the sanitarium's kitchen, and his body was found in the spa, confirming York's story about the casual worker he'd hired at Steeple Hill.

The second name to ring alarm bells was that of Carl Philips, a forty-six-year-old paranoid schizophrenic who had gone missing from a state psychiatric hospital more than a decade before. Not only were his remains the oldest that had been found at the sanitarium, but his grandfather had been the founder of Cedar Heights. Philips had inherited the derelict property but never bothered to develop it. It had lain fallow and forgotten, inhabited only by the termites and dragonflies.

Until Kyle had put it to use.

But it was the discovery of the third ID that caused most consternation. It belonged to a twenty-nine-year-old morgue assistant from Memphis, whose faded driver's licence was lying on the cabinet under the victims' photographs. His remains had been recovered from undergrowth by the pond and positively identified from dental records.

His name was Kyle Webster.

'He'd been dead eighteen months,' Jacobsen told me, when I'd called her after seeing a news report on TV. 'There're going to be questions about how an impostor could have secured a job in the morgue, but in fairness his documentation and references were authentic. And there was enough of a resemblance to the real Webster to fool anyone who only had old photographs to go on.'

I supposed it was in keeping with everything else he'd done. The man we'd known as Kyle Webster had delighted in misdirection all along. It shouldn't have come as a surprise that he'd slipped into the life of one of his victims as easily as he had the sloughed skin from their hands.

'So if he wasn't Kyle Webster, then who was he?' I asked.

'His real name was Wayne Peters. Thirty-one years old, from Knoxville originally, but worked as a morgue assistant in Nashville and then Sevierville, until he disappeared off the map two years ago. But it's his background before then that's interesting. Father unknown, mother died when he was an infant, so he was brought up by his aunt and uncle. Extremely bright by all accounts, did well at high school and even applied for medical college. Then things went sour. Around the time he was seventeen school records show he suddenly seemed to lose interest. He didn't make the grades he needed and wound up working for the family business until it went broke when his uncle died.'

'Family business?'

'His uncle owned a small slaughterhouse. They specialized in pork.'

I shut my eyes. *Pigs.*

'His aunt was his last remaining relative, and she died years ago,' Jacobsen went on. 'Natural causes, so far as we can tell. But you can probably guess where she and the uncle were buried.'

There was only one place, really.

Steeple Hill.

Jacobsen also gave me one other piece of information. When Wayne Peters's medical records were examined, it was found that as a teenager he'd had several operations to remove nasal polyps. They'd been successful, but the repeated cauterizations had resulted in a condition known as anosmia. Insignificant in itself, it answered the question Gardner had raised in the spa at Cedar Heights.

Wayne Peters had no sense of smell.

The recovery operation at the sanitarium was still going on, the grounds being dug up to ensure no more victims' remains were concealed. But my own role there had ended after that first day. By then not only had other faculty members from the Forensic Anthropology Center joined the effort, but the scale of the operation meant that the regional DMORT – Disaster Mortuary Operational Response Team – had also been called in. They'd arrived with a fully equipped portable morgue unit, and less than twenty-four hours after Paul and I had first climbed through its fence, the sanitarium and its grounds swarmed with activity.

I'd been politely thanked for my help and told I'd be contacted if my presence was required beyond the statement I'd already given. As I'd been driven through the ranks of TV and press vehicles camped beyond the sanitarium's gates, I'd felt both relief and regret. It felt wrong leaving an investigation like that, but then I reminded myself that it wasn't really my investigation.

It never had been.

I'd been prepared to either extend my stay in Tennessee for Tom's memorial service, or even fly back for it later if I had to. But in the end there had been no need. Regardless of what factors had contributed to it, Tom had died in hospital of natural causes, and so the formality of an inquest had been avoided. I was glad for Mary's sake, even though it left a sense of unfinished business. But then what death doesn't?

There had been no funeral. Tom had donated his body for medical research, though not at the facility. That would have been too disturbing for his colleagues. Mary had been dignified and dry-eyed at the service, standing beside a plump middle-aged man in an immaculate suit I didn't at first realize was their son. He carried himself with the faintly irritable air of a man who had better things to do, and when I was introduced to him afterwards his handshake was limp and grudging.

'You work in insurance, don't you?' I said.

'Actually, I'm an underwriter.' I wasn't sure what the distinction was but it didn't seem worth asking. I tried again.

'Are you staying in town long?'

He looked at his watch, frowning as though he were already late. 'No, I'm catching a flight back to New York this afternoon. I've had to reschedule meetings as it is. This came at a really bad time.'

I bit off the retort I'd been about to make, reminding myself that whatever else he might be, he was still Tom and Mary's son. As I walked away he was looking at his watch again.

Gardner and Jacobsen had both attended the ceremony. Jacobsen had returned to work already, the dressing on her shoulder all but invisible under her jacket. Gardner was still technically on sick leave. He'd suffered a transient ischaemic attack — a mini-stroke — from being held for so long in the chokehold. It had left him with slight aphasia and loss of sensation on one side, but only temporarily. When I saw him the only noticeable after-effect was a deepening of the corduroy-like lines in his face.

'I'm fine,' he told me, a little stiffly, when I asked how he was. 'There's no reason I can't work now. Damn doctors.'

Jacobsen looked as pristine and untouchable as ever. Except for slightly favouring her left arm no one would have known she'd been shot.

'I heard a rumour that she's up for a commendation,' I said to Gardner, while she was offering her condolences to Mary.

'It's under review.'

'For my money she deserves one.'

He unbent a little. 'Mine too, for what it's worth.'

I watched as Jacobsen spoke solemnly to Mary. The line of her throat was lovely. Gardner cleared his throat.

'Diane's still getting over a tough time. She broke up with her partner last year.'

It was the first hint of a personal life I'd had about her. I was surprised he'd offered the information.

'Was he a TBI agent as well?'

Gardner busied himself brushing something from the lapel of his creased jacket.

'No. She was a lawyer.'

Before they left, Jacobsen came over to say goodbye. Her grip was strong, the skin dry and warm as she shook my hand. The grey eyes seemed a little warmer than they had, but perhaps that was my imagination. The last I saw of her she was walking back to the car with Gardner, graceful and athletic beside the older agent's crumpled figure.

The ceremony itself was simple and moving. There had been no hymns, only two of Tom's favourite jazz tracks to start and close: Chet Baker's 'My Funny Valentine' and Brubeck's 'Take Five'. I'd smiled when I'd heard that. In between had been readings from friends and colleagues, but at one point the solemnity was broken by a baby's crying. Thomas Paul Avery howled lustily, despite his mother's best efforts to calm him.

No one minded.

He'd been born not long after Sam arrived at hospital, perfectly healthy and squalling his annoyance at the world. Sam's blood pressure had caused the doctors some concern at first, but it had returned to normal with remarkable speed after the birth. Within two days she'd been back at home, still pale and hollow-eyed when I'd visited, but with no other visible signs of her ordeal.

'It seems more like a bad dream than anything else,' she admitted, when Thomas had fallen asleep after nursing. 'It's like a curtain's been pulled across it. Paul's worried I'm in denial, but I'm not. It's more like what happened afterwards is more important, you know?' She'd been gazing down at her son's wrinkled pink face, but now she looked up at me with a smile so open it broke my heart. 'It's like all the bad doesn't matter. It's wiped everything else out.'

Of the two of them, Paul seemed to be finding it harder to deal with what had happened. In the days immediately afterwards, there was often a shadow in his face. It didn't take a psychologist to know he was reliving the ordeal, still cut by how close they'd come, and what might have been. But whenever he was with his wife and son the shadow would lift. It was still early days, but looking at the three of them together I felt sure the wounds would heal.

They usually do, given time.

My tea had gone cold. With a sigh I stood up and went to the phone to play back my messages.

'Dr Hunter, you don't know me, but I was given your number by DSI Wallace. My name is—'

The sound of the doorbell drowned out the rest. I paused the play-back and went to answer it. The last of the daylight filled the small entrance hall with a golden glow, like a forerunner of summer. I reached out to open the front door and was overwhelmed by a swooping sense of déjà vu. *A young woman in sunglasses stands outside in the sunlit evening. Her smile turns into a snarl as she reaches into her bag and pulls out the knife . . .*

I shook my head, scattering the images. Squaring my shoulders, I unlocked the front door and threw it wide open.

An elderly woman beamed up at me from the step. 'Ah, Dr Hunter, it is you! I heard someone moving about downstairs and wanted to make sure everything was OK.'

'Everything's fine, thanks, Mrs Katsoulis.' My neighbour lived in the flat above mine. I'd hardly spoken to her before I'd been attacked

316

the year before, but since then she'd taken it upon herself to turn vigilante. All four foot ten of her.

She hadn't finished with me yet. She peered past the hall into the living room, where my bags were still waiting to be unpacked.

'I thought I hadn't seen you around for a while. Have you been anywhere nice?'

She stared up at me expectantly. I felt my mouth start to twitch as I fought down an urge to laugh.

'Just a work trip,' I said. 'But I'm back now.'

Acknowledgements

Whispers of the Dead is a work of fiction, but the Anthropology Research Facility in Tennessee is real. Thanks are therefore due to Professor Richard Jantz, Director of the Forensic Anthropology Center in Knoxville, for granting permission to feature the facility and for his help with technical aspects. Dr Arpad Vass provided his usual quick responses to forensic queries and allowed Tom Lieberman to borrow his research, while Kristin Helm, Public Information Officer of the Tennessee Bureau of Investigation, was a mine of valuable information.

Thanks to my agents, Mic Cheetham and Simon Kavanagh, to Camilla Ferrier and all at the Marsh Agency, Simon Taylor and the team at Transworld, Caitlin Alexander at Bantam Dell, Peter Dench, Jeremy Freeston, Ben Steiner and SCF. I'd also like to thank my sister Julie and Jan Williams, without whom the writing of this book would have almost certainly have taken far longer: as someone who has now fully recovered from ME, I can recommend the Lightning Process to anyone who hasn't.

Finally, as ever, a huge thank you to my wife, Hilary. I really couldn't do it without her.

Simon Beckett, 2008